Wireless Networks

Series Editor

Xuemin Sherman Shen, University of Waterloo, Waterloo, ON, Canada

The purpose of Springer's Wireless Networks book series is to establish the state of the art and set the course for future research and development in wireless communication networks. The scope of this series includes not only all aspects of wireless networks (including cellular networks, WiFi, sensor networks, and vehicular networks), but related areas such as cloud computing and big data. The series serves as a central source of references for wireless networks research and development. It aims to publish thorough and cohesive overviews on specific topics in wireless networks, as well as works that are larger in scope than survey articles and that contain more detailed background information. The series also provides coverage of advanced and timely topics worthy of monographs, contributed volumes, textbooks and handbooks.

Zhongxiang Wei • Xu Zhu • Jingjing Wang •
Sumei Sun

Energy Efficient Full Duplex Wireless Communication Systems

Springer

Zhongxiang Wei
Department of Electrical Engineer and Electronics
Tongji University
Shanghai, China

Xu Zhu
School of Electronic and Information Engineering
Harbin Institute of Technology (Shenzhen)
Shenzhen, China

Jingjing Wang
School of Cyber Science and Technology
Beihang University
Beijing, China

Sumei Sun
Institute for Infocomm Research
Agency for Science Technology&Research
Singapore, Singapore

ISSN 2366-1186 ISSN 2366-1445 (electronic)
Wireless Networks
ISBN 978-981-97-8914-6 ISBN 978-981-97-8915-3 (eBook)
https://doi.org/10.1007/978-981-97-8915-3

This Springer imprint is published by the registered company Springer Nature Singapore Pte Ltd.
The registered company address is: 152 Beach Road, #21-01/04 Gateway East, Singapore 189721, Singapore

Preface

In an effort to overcome the spectral efficiency (SE) loss and high latency by half-duplex (HD), full-duplex (FD) has attracted extensive attention in industry and academia. With signal transmitted and received simultaneously over the same frequency, FD can theoretically double the SE over HD. More than that, by enabling the capability of simultaneous transmission and reception, FD becomes a key enabler for many promising 5G techniques, including but not limited to integrated sensing and communication, low-latency relaying, simultaneous bi-directional uplink and downlink transmission, etc. Nevertheless, in FD communications, both transmit- and receive-radio frequency (RF) chains are activated for exchanging data, and additional power is triggered by self-interference cancellation (SIC). As FD introduces additional power consumption, it is against the green evolution requirement proposed by the future communication systems. To address the critical high power challenge in applying FD communications, this book will introduce the fundamentals and state-of-the-art algorithm designs for energy-efficient FD designs.

In the first chapter, this book will present research background and fundamentals of FD communications. It will first present the classifications of FD techniques and demonstrate the SIC techniques for FD communications. Starting from the passive suppression in propagation domain, it will discuss the principles of analogue cancellation, which is an active cancellation technique in radio frequency domain. Despite the advances in propagation and analog domain SIC, digital domain solutions for SIC is still needed. Hence, this book will then discuss the digital domain SIC schemes. Finally, it will review the power consumption of FD communications, which motivates the energy efficiency (EE)-oriented FD designs.

In Chap. 2, to strike the balance between system SE and power consumption, a hybrid duplexing strategy is discussed for distributed antenna (DA) systems, where antennas are enabled to work in hybrid FD, HD, and sleeping modes. To maximize the system EE with a reasonable level of computational complexity, a channel gain-based DA clustering algorithm is reviewed for the activation/deactivation of transmit- and receive-chains, which highlights the characteristics of distributed deployment of antennas. Based on the DAs' configuration, a distributed hybrid duplexing-based and EE-oriented algorithm is discussed to further optimize the downlink beamformer and the uplink transmission power. In particular, self-interference at DAs, co-channel interference from uplink users to downlink users, and multiuser interference in both uplink and downlink are considered. Performance results show that the hybrid duplexing system provides both EE and SE enhancements than the co-located FD systems, showing significant advantages in alleviating high path loss as well as in cutting the power consumption.

In Chap. 3, this book will discuss the application of FD in wireless power transfer (WPT) relaying systems. Due to the limited transmission power at the WPT relay, SE of source-relay link becomes a bottleneck in system end-to-end SE performance. It will first review relaying protocols for downlink orthogonal frequency division multiple access (OFDMA) systems, in which a WPT FD relay operates in decode-and-forward (DF) mode. In particular, the time slot durations of the source-relay and relay-users hops are designed to be asymmetric and dynamic, in order to improve the degrees of freedom and hence the system SE. Capturing the non-linearity of harvesting efficiency of the relay node, a near optimal iterative algorithm is discussed by jointly optimizing the time slot lengths, subcarriers allocation, and transmission power at the source and the relay. Furthermore, self-interference generated at the relay is utilized as a constructive energy source rather than being mitigated, which increases substantially the amount of energy harvested at the FD relay.

In Chap. 4, this book will discuss the application of FD relaying in nonorthogonal multiple access (NOMA) systems. The FD cooperative NOMA achieves a higher throughput over the traditional HD cooperative NOMA, where the strong users (SUs) with good channel conditions can act as an FD relay node for the weak users with poor channel conditions. However, the EE of a cooperative NOMA may be degraded due to the additional power consumption incurred at the SUs side. Therefore, this chapter will discuss the EE maximization design of an FD cooperative NOMA system. Starting from an interesting "signal-to-inference-noise ratio gap reversal" problem of cooperative NOMA systems, it will explain the challenges of successive interference cancellation performance degradation at the SUs. Then, it will review a low complexity algorithm for maximizing system's EE while guaranteeing a successful successive interference cancellation operation. Finally, the performance results demonstrate that the algorithm achieves both a higher EE and throughput over the classic HD cooperative NOMA and non-adaptive FD cooperative NOMA.

Finally, we would like to thank audience, authors, and reviewers that have contributed to this book. We sincerely hope that students, engineers, and researchers enjoy the contributions and the compilation proposed by this book.

Shanghai, China — Zhongxiang Wei
Shenzhen, China — Xu Zhu
Beijing, China — Jingjing Wang
Singapore, Singapore — Sumei Sun

Contents

List of Figures

List of Tables

Chapter 1
Fundamentals of Full Duplex Communications

In this chapter, we will briefly present research background of full-duplex (FD) communications. Then, we will present the technical fundamentals of FD techniques.

1.1 Background of Full Duplex Communications

To date, five generations of cellular communication systems have been adopted in the world with each new mobile generation emerging every 10 years or so. First generation analogue cellular systems in 1981; second generation digital technology in 1992, third generation (CDMA 2000, WiMAX) in 2001, fourth generation (LTE, LTE-A) in 2011, and fifth generation (5G NR) in 2018. As the demand for throughput in mobile broadband communications increases exponentially every year, it is estimated that the end-user traffic demand will increase at a compounded annual growth rate of 22%, from 2020 to 2030. Global network traffic is expected to reach between 2443 to 3109 exabytes per month in 2030, requiring researchers and engineers to seek greater capacity to support richer multimedia services [1]. 5G-NR specifies three major scenarios for communications, i.e., enhanced mobile broadband (eMBB), ultra reliable and low latency communications (uRLLC), and massive machine type communications (mMTC). In eMBB, higher throughput is specified. For the downlink, up to 50 Mbps are offered for outdoor and 1 Gbps for indoor, with half of these values available for the uplink. In uRLLC, extremely high reliability and low latency are expected. For instance, for remote control of process automation, a reliability of 99.9999% is required, with a user experienced throughput up to 100 Mbps and an end-to-end latency of 50 ms. In mMTC, it requires the 5G systems to support very high traffic densities of devices, as typically summarized in Table 1.1.

To support the enhanced system metrics, advanced techniques have been developed and standardized as key enablers in 5G communications, e.g., full dupelx (FD),

Z. Wei et al., *Energy Efficient Full Duplex Wireless Communication Systems*, Wireless Networks, https://doi.org/10.1007/978-981-97-8915-3_1

Table 1.1 Requirements and realities of 1G through 5G cellular systems

Generation	Requirements	Comments
1G	No official requirements. Analogue technology	Deployed in 1980
2G	No official requirements Digital technology	First digital systems New service such as SMS and low-rate data
3G	ITU's IMT-2000 required 144 kbps mobiles, 2 Mbps indoor	Primary technologies include CDMA2000 1×EV-DO WiMAX
4G	ITU's IMT advanced Requirements include ability to operate in up to 49 MHz radio channels and with high SE ability	IEEE 802.16m LTE-Advanced
5G	For peak data rate: Downlink: 20 Gbit/s, Uplink: 10 Gbit/s For peak spectral efficiencies: Downlink: 30 bit/s/Hz, Uplink: 15 bit/s/Hz User plane latency (single user, small packets): 4 ms for eMBB, 1 ms for URLLC control plane latency (idle to active): 10–20 ms maximum aggregated system bandwidth: at least 100 MHz, up to 1 GHz in higher frequency bands (above 6GHz) mobility: up to 500 km/h in rural eMBB	3GPP R15-R19

massive multiple-input and multiple-output (MIMO), millimeter (mm) wave, etc. For example, classic communication systems work in half-duplex (HD) mode [2], where antennas receive and transmit signal in different time slots or orthogonal frequencies. This leads to loss of spectral efficiency (SE). As a contrast, FD technique can receive and transmit signal simultaneously at the same frequency, and there provides SE enhancement over existing HD systems. However, strong self-interference is introduced at FD node's receiver from its transmitter, which requires strict self-interference cancellation (SIC) operation for FD nodes. According to recent researches, SIC amount of more than 100 dB can be achieved [3], paving the way for the commercial use of FD techniques.

To implement FD in 5G communications, additional power consumption by the SIC operation is a critical challenge [4]. Moreover, for a HD node, only one transmit-chain is activated for signal transmission (or one receive-chain is activated for signal reception) in each time slot. While for an FD node, its receive- and transmit- chains are always activated, which apparently increases the power consumption of FD systems [5]. In recent years, energy efficient cellular networks are gaining importance due to the severity of carbon emissions levels in air. The increasing energy consumption of mobile networks increases the operator's costs as well. The ecological concerns have led to the evolution of green communication in cellular networks. As a consequence, the advent of 5G-NR is expected to increase energy efficiency (EE) up to 1000x, with service provisioning to billions of devices. To address the high power consumption issue, EE is defined as the ratio between system SE and total power consumption, which becomes an important system performance metric proposed in 5G-NR.

1.2 Fundamentals of FD Communications

In this section, we will discuss the fundamentals of FD communications, from perspectives of classification, SIC operation and power consumption.

1.2.1 FD Classifications

We now discuss the classifications of FD systems.

By with or Without Relay Without the deployment of relay nodes, FD systems often denote the communication scenario between two FD nodes. As illustrated in Fig. 1.1, two nodes can transmit signals to each other simultaneously at the same frequency, and this bi-directional transmission requires SIC operations at both two nodes. In long-distance communications, a relay node may be required to assist data

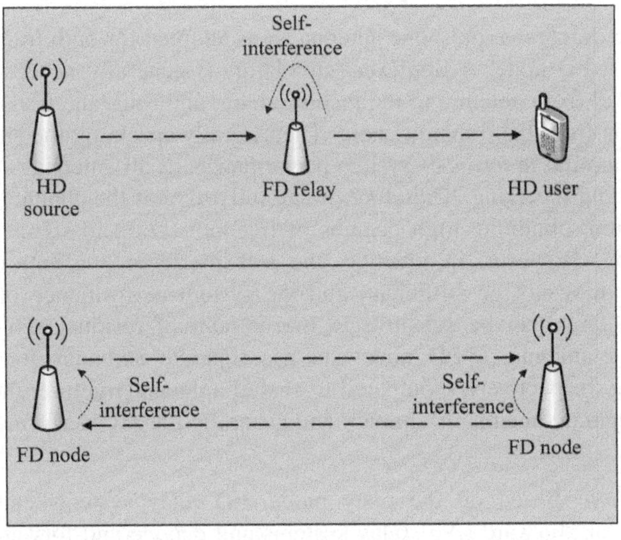

Fig. 1.1 FD relaying systems and MIMO point-to-point FD

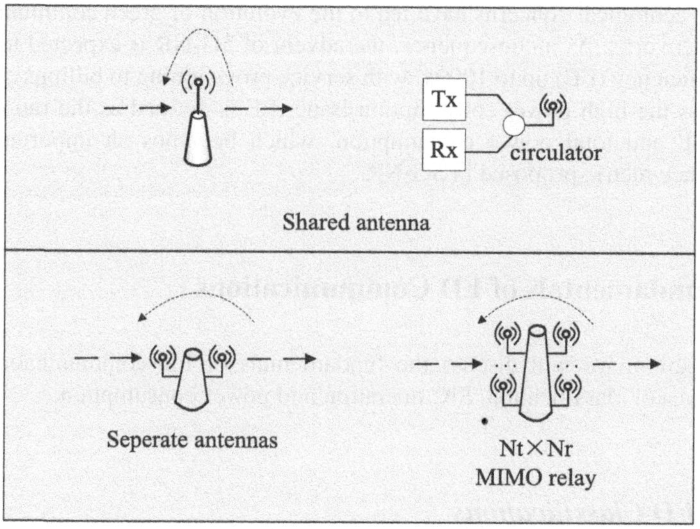

Fig. 1.2 Shard and separated antenna FD relaying systems

exchange between source and destination. If the relay node works in FD mode, SIC is required at the relay node. Also, bi-directional relay is another FD applications, where two nodes exchange their data aided by an FD relay. In this case, the two nodes and the relay all work in FD mode, where SIC is needed for the three nodes.

By Antenna Type According to the type of antennas, FD systems can be classified as shared-antenna FD [6] and separate-antenna FD [2], as illustrated by Fig. 1.2.

With shared-antenna, only one antenna set is adopted for both transmission and reception at a FD node. A duplexer (circulator) is generally needed to route the received signal from antenna to the receive chain, and route the transmitted signal to the antenna from the transmit chain. Theoretically speaking, the shared-antenna structure is optimal in terms of system performance, as all antennas are capable of transmitting and receiving. Therefore, it can fully exploit the channel diversity for communication, obtaining high degrees of freedom (DoF) in signal transmission and reception. However, in practice, the self-interference isolation offered by the duplexer may not be sufficient, and the system performance of the shared-antenna structure may be sensitive to the amount of residual self-interference. With separate-antenna, a FD node can use separate antennas for transmitting and receiving, respectively. Compared to shared-antenna structure, the SIC circuit design for separate-antenna is simpler, and generally the SIC performance is easier to guarantee.

By Relay Mode Based on the relay mode, FD relay systems can be divided into amplify-and-forward (AF) relay systems and decode-and-forward (DF) relay systems [2, 4]. For an AF relay node, it subtracts the self-interference, amplifies the

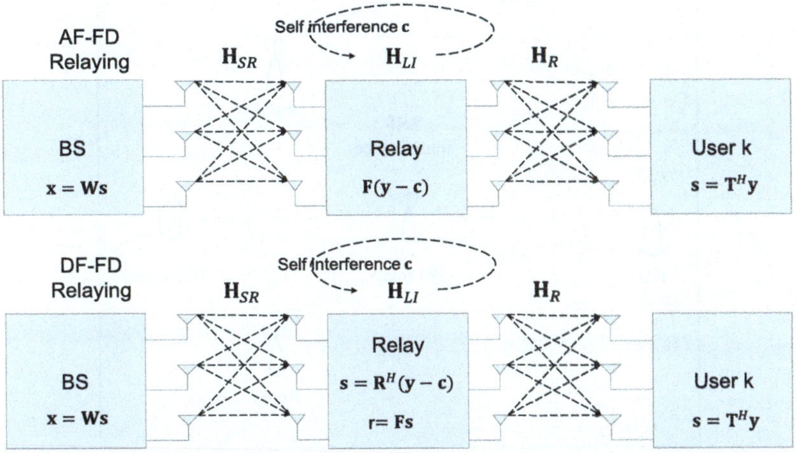

Fig. 1.3 AF and DF FD relaying systems

received signal from source, and finally forwards the signal (including the residual self-interference and noise) to destination. By contrast, a DF relay node performs SIC, decodes the received signal, re-encodes the signal, and finally transmits the signal to its destination. Indeed, as an AF FD relay node does not need to decode signal, it generally provides low latency transmission and low power consumption compared to an DF FD node. However, as the noise and residual self-interference is amplified and forwarded to destination, the end-to-end SINR may be lower than that of DF relaying systems, as illustrated in Fig. 1.3.

By Numbers of Relays or Users As illustrated by Figs. 1.4 and 1.5, a pair of communication nodes may be aided by multiple relay nodes [7]. In this case, it becomes possible to select subset of the relays for forwarding signal and exploit the diversity in relay node selection. Also, multiple users may be served by one relay node. In order to manage the multiuser interference, orthogonal frequency division duplexing access (OFDMA) can be employ, where orthogonal subcarriers are allocated for different users [2]. In this case, per-subcarrier SIC may be desirable to guarantee a reasonable communication performance, which however requires complicated SIC circuit design.

By Transmission Directionality According to the directionality of data exchange, it can be classified into one-way FD and two-way FD system [8], as demonstrated by Fig. 1.6. One-way FD system denotes that an FD relay node receives and forwards signal in only one direction, while two-way systems FD denotes that an FD relay assists data exchange between a pair of nodes. Generally speaking, two-way FD systems can provide higher overall throughput, which however needs more complicated SIC operations.

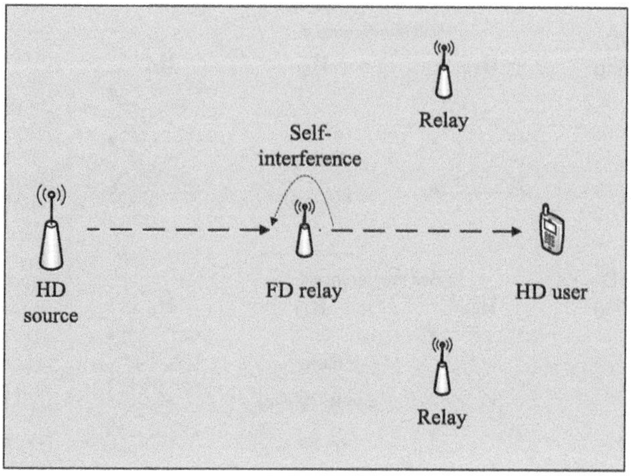

Fig. 1.4 FD relaying system with multiple FD relay nodes

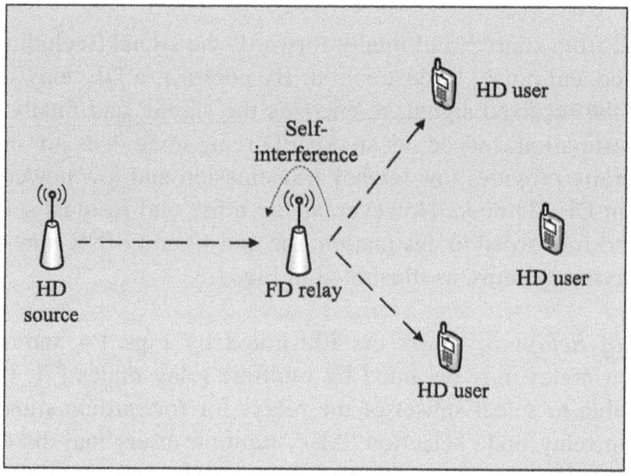

Fig. 1.5 Multiuser FD relaying systems

Mimicking FD Systems by HD Operations There are some schemes using HD systems to mimic FD systems, as illustrated by Fig. 1.7. One approach is to use two buffer-aided HD relays to assist communication between source and destination, where one relay receives the signal from source and the other transmits the buffered signal to destination [9]. However, as two HD relays are required at the same time, the incurred power consumption is also high. Another approach is to let an HD relay node receive and transmit at different frequencies, which is sometimes called out-band FD transmission. In this case, as the whole frequency bandwidth is separated for orthogonal transmission and reception, the SE is decreased compared to FD relaying systems.

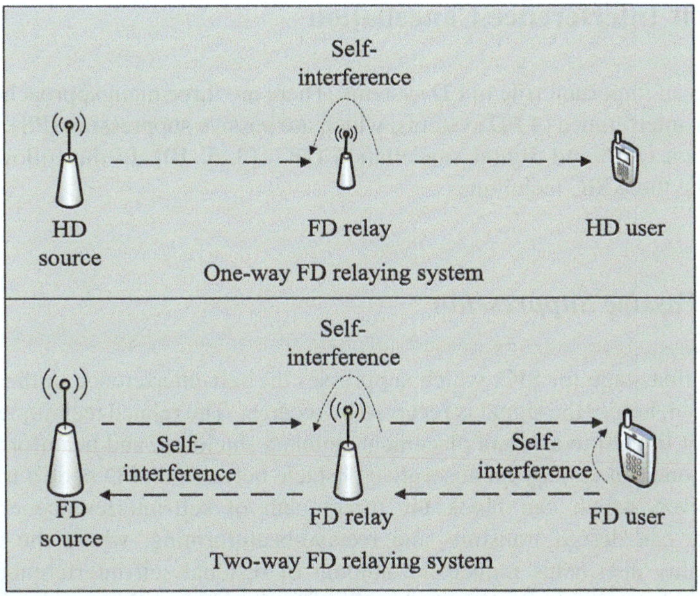

Fig. 1.6 One-way and tow-way FD relaying systems

Fig. 1.7 Mimicking FD relaying with two buffer-aided HD relays and out-band FD relaying

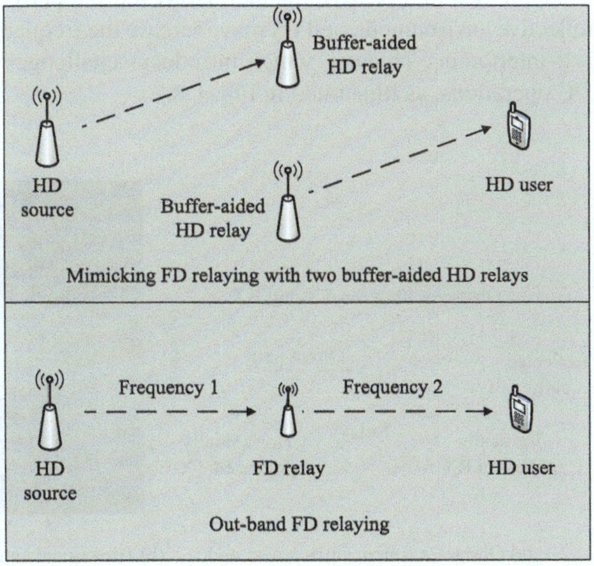

1.3 Self-Interference Cancellation

SIC plays an important role in FD systems. There are three main approaches to suppress self-interference of FD systems, which are passive suppression (PS), analogue cancellation (AC) and digital cancellation (DC) [3, 5, 10]. In the followings, we discuss the three SIC techniques.

1.3.1 Passive Suppression

PS is the first stage for SIC, which suppresses the self-interference in the propagation domain, before the signal is received by receiver. The related techniques include but are not limited to antenna placement, antenna shielding, and beamforming. For example, one can employ an absorptive obstacle between an FD node's transmitter and receiver, which can block the direct path of self-interference effectively. Also, one can design transmit- and receive-beamforming, where the enhanced directionality also helps reduce the amount of residual self-interference. Recent researches demonstrate that more than 70 dB of self-cancellation amount can be obtained by PS in an anechoic chamber, and more than 40 dB in a reflected room [11]. The advantage of PS lies in its low power consumption, where self-interference is suppressed in a passive manner. However, the performance of PS is sensitive in reflective environment, and PS may increase the frequency selectivity of the residual self-interference channel, which introduces challenges for the subsequent AC and DC operations, as illustrated in Fig. 1.8.

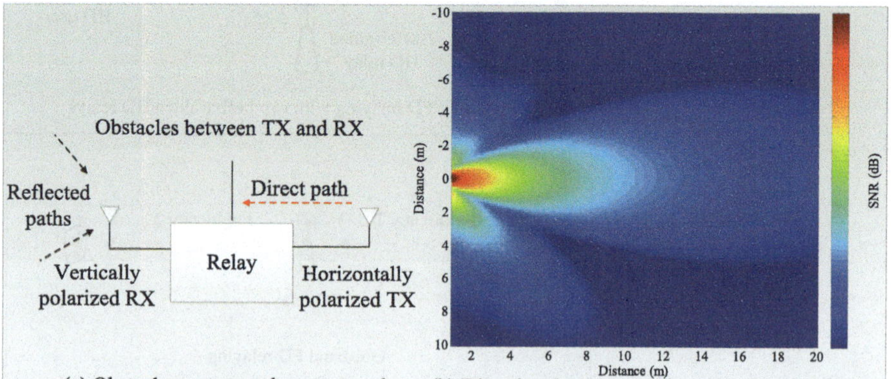

(a) Obstacles, antenna placement and cross polarization (b) Directional antenna with narrow beamwidth at 60 GHz FD relay

Fig. 1.8 Passive suppression for a FD relay node

1.3.2 Analogue Cancellation

After signal is received by receiver, self-interference can be further suppressed by AC operation, before signal goes through low noise amplifiers (LNA)s [5, 12, 13]. There are two kinds of AC designs: direct-conversion architecture AC and non-direct-conversion architecture AC. The first architecture estimates self-interference and subtracts it in the analogue domain directly. As it does not require additional baseband signal processing unit, the power consumption is lower than the non-direct-conversion architecture AC. Differently, the non-direct-conversion AC architecture introduces the predicted self-interference in digital domain, converts the digital signal to analogue, and finally feeds it to the receive chain for AC operation. Since baseband signal processing unit, digital-to-analogue converters (DACs), mixers, low pass filters (LPFs), attenuators, and adders are needed. The incurred power consumption by the non-direct-conversion AC architecture is as high as the equivalent transmit-chains, as illustrated in Fig. 1.9.

1.3.3 Digital Cancellation

DC is the last stage for SIC, which subtracts the residual self-interference after PS and AC in digital domain [10]. In order to well suppress the residual self-interference, DC needs to estimate the residual self-interference, and subtract it accurately. Therefore, complicated baseband signal processing unit is required by DC operation, which consumes non-negligible power consumption as shown in Fig. 1.10. The summary of different SIC in FD systems is given by Table 1.2.

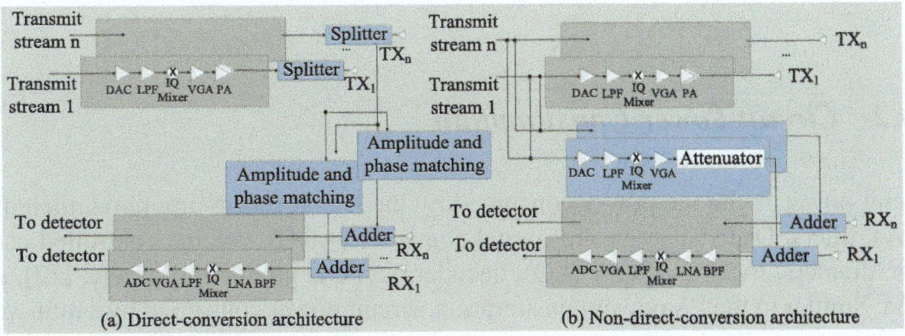

Fig. 1.9 Analog cancellation for FD, with direct-conversion and non direct-conversion architectures

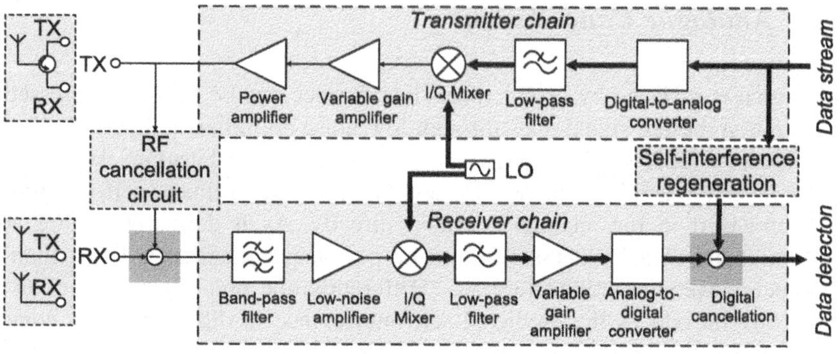

Fig. 1.10 Digital cancellation operation suppresses self-interference in digital domain

1.4 Power Consumption and EE Challenges in FD Systems

In this section, we discuss power consumption of FD systems.

1.4.1 Power Amplify Power Consumption

For wireless communication systems, the power consumed by cascaded power amplifiers (PA)s contributes a large part to the overall power consumption, which is mainly caused by the low drain efficiency. For example, the drain efficiency of PAs working at 2.4/5 GHz is generally on the level of 30–40% [14, 15], while for PAs working at mm-wave frequency, the drain efficiency is generally lower than 25% [16].

1.4.2 Circuit Power Consumption

Circuit power is the the power consumed by the signal processing parts, including LNAs, filters, DAC, analogue-to-digital converter (ADC), voltage controlled oscillator, dividers, local oscillator (LO), decoder, variable gain amplifier etc. [17], and so on. Similar to the PA power consumption, circuit power consumed by a mm-wave chip is generally higher than that consumed by a cm-wave chip [17].

Table 1.2 Summary of different SIC schemes for FD systems in terms of EE

SIC schemes	Approaches	Additional power consumption	Features	Drawbacks
PS	Benefiting from high PL between relay's transmitter and receiver	Nil	Much higher PL than the PL with cm wave	Sensitive to environment; Increased frequency selectivity of self-interference channel
	Antenna shielding	Nil	Fully exploiting the advantage of mm level wavelength	Sensitive to environment; Increased frequency selectivity of self-interference channel
	Directional antenna	Nil	Widely used in mm-wave communications; Relay's transmitter pointing to destination with no direct self-interference	Sensitive to environment; Increased frequency selectivity of self-interference channel
AC	Direct-conversion architecture	Low	Aware to the reflected self-interference	Sensitive to wideband self-interference
	Non-Direct-conversion architecture	High	Unaware to the reflected self-interference	Not feasible in AF relay systems
DC	Digital-domain canceller	High	Inefficient given good performance by PS+AC, may degrade the EE performance	May cause negative effect to system (the introduced noise power is higher than the power of the self-interference cancelled)

1.4.3 Additional Power Consumption Incurred by FD Transmission

As discussed in this chapter, HD mode transmits and receives signal in orthogonal time slots, and thus transmit-chains or receive-chains are activated in different time slots. By contrast, both transmit-chains and receive-chains are always activated in

FD systems, where the power consumption of FD mode is naturally higher than that of HD mode. Another additional power consumption of FD node is introduced by SIC operation. Generally speaking, complicated SIC requires more components and thus consumes higher power. For example, the AC operation design in [3] needs DAC, transmit radio unit and adder to mitigate self-interference. Another kind of AC design in [5] uses tunable attenuation and delay unit to route the estimated signal to the receiver for SIC, where the incurred power consumption is still non-negligible. Besides, since DC design needs to calculate residual self-interference after PS and AC operations, the power consumed by DC operation is also high [3].

Chapter 2
Energy Efficient Bi-Directional FD Distributed Antenna Systems

Conventional HD MIMO splits uplink and downlink transmission into orthogonal time slots or frequencies [18, 19]. FD MIMO system allows communication nodes to transmit and receive signals simultaneously over the same frequency [20], subject to a good performance of SIC [11, 21, 22]. SE maximization design has been extensively investigated to exploit the advantage of FD MIMO systems [10, 23–25]. In [23], an algorithm was proposed to optimize the system ergodic mutual information. In [24, 25], the SE optimization of bidirectional FD MIMO systems were investigated, where a FD BS communicates with multiple single-antenna users in downlink and uplink simultaneously. In [10], a multiuser FD MIMO precoding applicable for single-carrier and OFDM systems was designed to maximize system SE. More literature review of bi-directional FD MIMO systems is summarized in Table 2.1.

Some fundamental challenges need to be considered in FD MIMO systems. (a) MIMO increases the number of active transmit/receive chains, leading to much higher power consumption than that of single-input and single-output (SISO) communications [18]. (b) For FD systems, additional power consumption is incurred by SIC operation [4, 5]. (c) In bi-directional FD MIMO communications, there are self-interference from the BS's transmitter to its receiver, co-channel interference from uplink users to downlink users, and downlink multiuser self-interference. Hence, interference management is critical in guaranteeing the performance of bi-directional FD MIMO communications. (d) Classic MIMO systems deploy antennas in a centralized manner, where SIC operation becomes difficult with an increasing number of antennas. Due to the above challenges, though there are some works considering energy efficient design for HD one-directional transmission systems [26–28], they may not be directly applied to bi-directional FD systems.

In recent years, distributed MIMO has attracted much attention with some potential advantages. (a) In practice, users are separated at different locations [36]. By deploying distributed antennas (DA), it is possible to only activate the DAs close to users, thereby significantly reducing system power consumption. (b) In

Z. Wei et al., *Energy Efficient Full Duplex Wireless Communication Systems*, Wireless Networks, https://doi.org/10.1007/978-981-97-8915-3_2

Table 2.1 Summary of MIMO FD networks

Reference	Precoding or decoding method used	Joint or separate beamforming design	Optimisation metric	Antenna array
[2]	ZF	Joint	Throughput maximisation	Co-located
[29]	ZF	Joint	Outage probability minimisation	Co-located
[30]	MMSE	Separate	Bit error rate minimisation	Co-located
[25]	\	Joint	Throughput maximisation	Co-located
[24]	MMSE	Joint	Throughput maximisation	Co-located
[31]	ZF	Joint	Outage probability minimisation	Co-located
[32]	ZF	Separate	Throughput maximisation	Co-located
[33]	ZF	Joint	Outage probability minimisation	Co-located
[34]	ZF	Joint	Throughput maximisation	Co-located
[35]	ZF	Separate	SIC maximisation	Co-located

distributed MIMO, antennas are separately deployed, and this natural isolation brings advantages for PS of SIC operation. (c) As antennas are closer to users, distributed MIMO is able to reduce path loss (PL) impact and obtain blockage-free effect [37], which also extends coverage area of communication systems. Due to the above advantages, distributed FD MIMO system was researched in [38], where the antennas can either work in FD or sleep mode [39]. Having only these two modes may not be optimal in practice. For example, the active antennas may be only close to uplink users while are far from downlink users. Hence, these antennas can work in HD mode, which provides communication services for signal transmission or reception, and at the same time incurs lower power consumption.

2.1 System Model for Bi-Directional FD DA Systems

2.1.1 System Model

Assume there is a central baseband processing unit, which connects L DAs. The DAs can work in FD, HD (only transmit or receive) or sleep mode, while the multiple single-antenna users only work in HD for low hardware complexity. Without loss of generality, there are U uplink users and K downlink users. Assume that the channel state information is obtained perfectly [37, 38]. The system model is illustrated by Fig. 2.1.

Define a receive chain activation vector $\boldsymbol{r} \in \mathbb{C}^{L\times 1}$, whose l-th element r_l is equal to 1 if the l-th DA's receive chain is activated, otherwise r_l is equal to 0. Similarly,

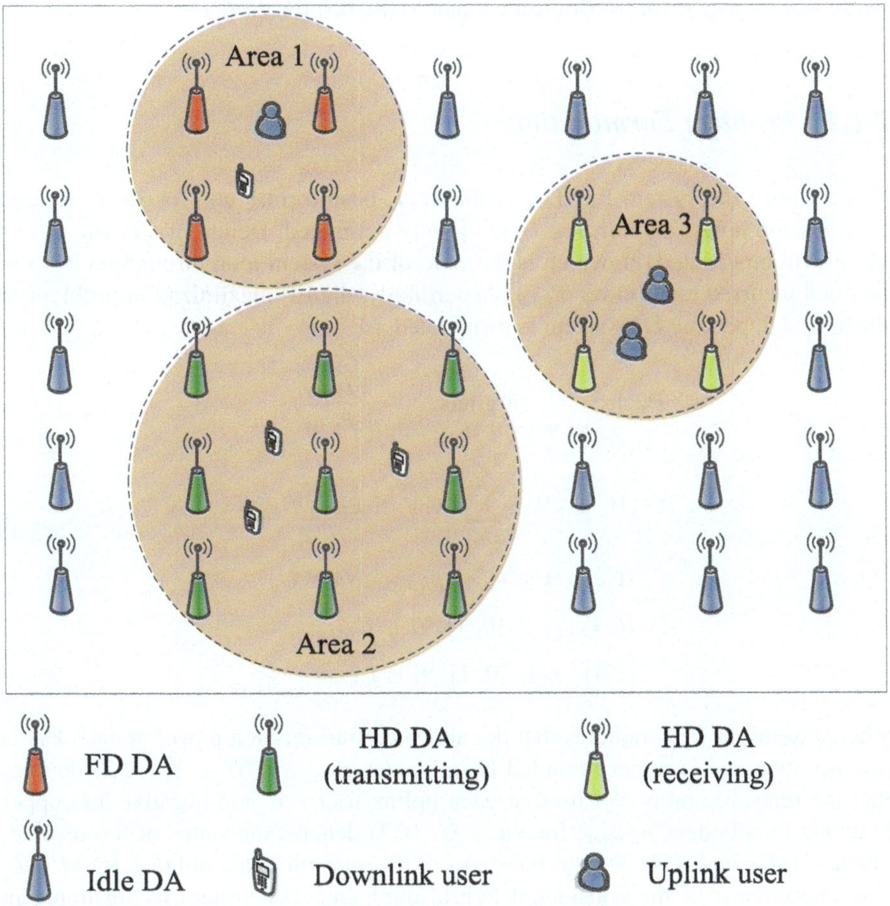

Fig. 2.1 Bidirectional DA systems, where DAs can work in FD, HD and sleeping modes

define a transmit chain activation vector $\boldsymbol{t} \in \mathbb{C}^{L\times 1}$, whose l-th element t_l equals to 1 if the l-th DA's transmit chain is activated, otherwise $t_l = 0$. Accordingly, there are four states for each DA, which can be described as 4 tuples. (0, 0) means both receive- and transmit-chains are deactivated, and thus the DA is in sleep mode. (1, 0) means only receive-chain is active while transmit-chain is deactivated. (0, 1) means only transmit-chain is active while receive-chain is deactivated. The above two cases mean that the DA works in HD mode. (1, 1) means both receive- and transmit-chains are activated and thus the DA works in FD mode. Since the shared-antenna deployment has been extensively researched, where only one antenna set is adopted for simultaneous transmission and reception. The received and transmitted signal are separated with the help of a duplexer [40], the configuration is reasonable in practice. Define $\boldsymbol{w_k} \in \mathbb{C}^{L\times 1}$ as downlink beamformer at DAs for transmitting signal to multiple users $\forall k \in K$. For each vector $\boldsymbol{w_k}$, its element w_k^l denotes the beamforming weight of antenna l for user k. Denote p_u as transmission power used by the uplink user u for transmitting signal to the BS, for $\forall u \in U$.

2.1.2 Problem Formulation

To maximize the system EE, DA vectors $\boldsymbol{r}, \boldsymbol{t}$, beamformer $\boldsymbol{w_k}$, for $\forall k \in K$, and transmission power p_u, for $u \in U$ are jointly optimized. Define $\eta(\boldsymbol{r}, \boldsymbol{t}, \boldsymbol{w_k}, p_u)$ as the EE (in bits/Joule/Hz), which is the ratio of the system total throughput T_{total} to the total incurred total power P_{total}. Accordingly, the EE maximization problem of the hybrid duplexing DA system is formulated as

$$\begin{aligned}
P2.1: \quad & \underset{\boldsymbol{r},\boldsymbol{t},\boldsymbol{w_k},p_u,k\in K,u\in U,l\in N}{\text{argmax}} \quad \frac{T_{\text{total}}}{P_{\text{total}}}, \\
s.t.\ (C1) &: 0 \le \sum_{k=1}^{K} |w_k^l|^2 \le p_{DA}, \forall l \in L, \\
(C2) &: 0 \le p_u \le p_{u,max}, \forall u \in U, \\
(C3) &: r_l = \{0, 1\}, \forall l \in L, \\
(C4) &: t_l = \{0, 1\}, \forall l \in L,
\end{aligned} \tag{2.1}$$

where constraint $(C1)$ denotes that the allocated transmission power at each DA is non-negative and is upper bounded by a budget p_{DA}, for $\forall l \in L$. $(C2)$ denotes that the transmission power used at each uplink user u is non-negative and upper bounded by a budget $p_{u,max}$, for $\forall u \in U$. $(C3)$ denotes the status of the receive-chain of DA l, and $(C4)$ denotes the status of the transmit-chain of DA l, for $\forall l \in L$. The interference of the considered hybrid duplexing DA system is illustrated in Fig. 2.2.

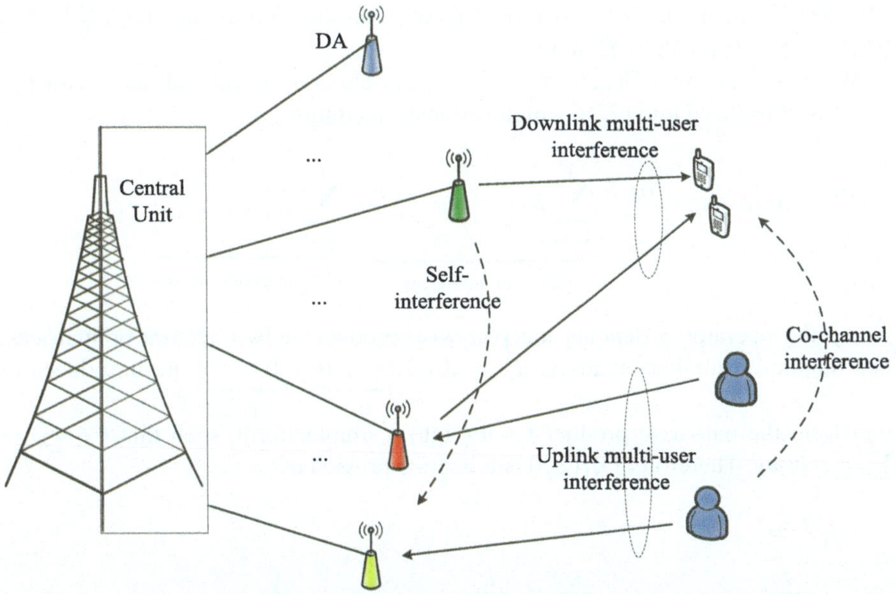

Fig. 2.2 Self-interference, downlink multi-user interference, co-channel interference, and uplink interference need to be suppressed for maintaining good communication performance

2.2 Throughput and Power Consumption Analysis

The optimization problem involves with throughput and power consumption in both downlink and uplink. In this section, the downlink and uplink throughput are analyzed respectively, and then the total power consumption is given.

2.2.1 *Downlink Throughput*

In each time slot, the received signal at downlink user k is given by

$$y_k^{DL} = \boldsymbol{h_k w_k} d_k^{DL} + \underbrace{\sum_{k' \neq k}^{K} \boldsymbol{h_k w_{k'}} d_{k'}^{DL}}_{\text{multiuser interference}} + \underbrace{\sum_{u=1}^{U} \sqrt{p_u} e_{u,k} d_u^{UL}}_{\text{co-channel interference}} + z_k, \tag{2.2}$$

where vector $\boldsymbol{h_k} \in \mathbb{C}^{1\times L}$ is downlink channel between L DAs and downlink user k, which includes the PL and small scale fading. In particular, the l-th element $(\boldsymbol{h_k})_l$ presents the channel condition between the l-th DA and downlink user k. $d_k^{DL} \in \mathbb{C}$ denotes the intended data for downlink user k. $e_{u,k} \in \mathbb{C}$ denotes the channel condition from uplink user u to downlink user k. $d_u^{UL} \in \mathbb{C}$ is transmit data from uplink user u. $z_k \sim CN(0, \sigma_k^2)$ is the complex Additive While Gaussian Noise

(AGWN) at downlink user k. Without loss of generality, it is assumed $\mathbb{E}\{|d_k^{DL}|^2\} = \mathbb{E}\{|d_u^{UL}|^2\} = 1$, for $\forall k \in K$ and $\forall u \in U$.

When the activation/deactivation of transmit-chain is considered, the vector $\boldsymbol{t} \in \mathbb{C}^{L\times 1}$ is introduced into (2.2), which is transformed into

$$y_k^{DL} = \boldsymbol{h}_k(\boldsymbol{t}\circ\boldsymbol{w}_k)d_k^{DL} + \underbrace{\sum_{k'\neq k}^{K} \boldsymbol{h}_k(\boldsymbol{t}\circ\boldsymbol{w}_{k'})d_{k'}^{DL}}_{\text{multiuser interference}} + \underbrace{\sum_{u=1}^{U}\sqrt{p_u}e_{u,k}d_u^{UL}}_{\text{co-channel interference}} + z_k, \tag{2.3}$$

where the operator $\circ$ denotes the pair-wise product of two vectors or matrices. For simplicity, auxiliary matrices $\boldsymbol{F}_l = diag\{\underbrace{0,\ldots,0}_{l-1}, 1, \underbrace{0,\ldots,0}_{L-l}\}$ are introduced to transform the pair-wise product $\boldsymbol{t}\circ\boldsymbol{w}_k$ into a simpler form, such that $\boldsymbol{t}\circ\boldsymbol{w}_k = \sum_{l=1}^{L} t_l \boldsymbol{F}_l \boldsymbol{w}_k$. Therefore, Eq. (2.3) can be re-expressed as

$$y_k^{DL} = \\ \boldsymbol{h}_k(\sum_{l=1}^{L} t_l\boldsymbol{F}_l\boldsymbol{w}_k)d_k^{DL} + \underbrace{\sum_{k'\neq k}^{K}\boldsymbol{h}_k(\sum_{l=1}^{L} t_l\boldsymbol{F}_l\boldsymbol{w}_{k'})d_{k'}^{DL}}_{\text{multiuser interference}} + \underbrace{\sum_{u=1}^{U}\sqrt{p_u}e_{u,k}d_u^{UL}}_{\text{co-channel interference}} + z_k. \tag{2.4}$$

Based on (2.4), the SINR of downlink user k can be calculated as

$$\Gamma_k^{DL} = \frac{\left|\boldsymbol{h}_k(\sum_{l=1}^{L} t_l\boldsymbol{F}_l\boldsymbol{w}_k)\right|^2}{\sum_{k'\neq k}^{K}\left|\boldsymbol{h}_k(\sum_{l=1}^{L} t_l\boldsymbol{F}_l\boldsymbol{w}_{k'})\right|^2 + \sum_{u=1}^{U} p_u\left|e_{u,k}\right|^2 + \sigma_k^2}. \tag{2.5}$$

The total downlink throughput is calculated as

$$T_{DL} = \sum_{k=1}^{K}\log_2(1+\Gamma_k^{DL}). \tag{2.6}$$

To simplify the mathematical expression, (2.6) is transformed into the structure $\log(1+\frac{A}{B}) = \log(A+B) - \log B$, such that

$$T_{DL} = \sum_{k=1}^{K}\log_2(\sum_{k'=1}^{K}\left|\boldsymbol{h}_k(\sum_{l=1}^{L} t_l\boldsymbol{F}_l\boldsymbol{w}_{k'})\right|^2 + \sum_{u=1}^{U} p_u|e_{u,k}|^2 + \sigma_k^2) - \\ \sum_{k=1}^{K}\log_2(\sum_{k'\neq k}^{K}\left|\boldsymbol{h}_k(\sum_{l=1}^{L} t_l\boldsymbol{F}_l\boldsymbol{w}_{k'})\right|^2 + \sum_{u=1}^{U} p_u|e_{u,k}|^2 + \sigma_k^2). \tag{2.7}$$

Define matrices $\boldsymbol{H}_k = \boldsymbol{h}_k{}^H \boldsymbol{h}_k$ and $\boldsymbol{W}_k = \boldsymbol{w}_k \boldsymbol{w}_k^H$. Then, the overall downlink throughput is calculated by (2.8).

$$T_{DL} = \sum_{k=1}^{K} \log_2\Big(\sum_{k'=1}^{K} \mathrm{Tr}\Big(\sum_{m=1}^{L}\sum_{n=1}^{L} t_m t_n \boldsymbol{F}_m \boldsymbol{W}_{k'} \boldsymbol{F}_n{}^H \boldsymbol{H}_k\Big) + \sum_{u=1}^{U} p_u |e_{u,k}|^2 + \sigma_k^2\Big)$$
$$- \sum_{k=1}^{K} \log_2\Big(\sum_{k' \neq k}^{K} \mathrm{Tr}\Big(\sum_{m=1}^{L}\sum_{n=1}^{L} t_m t_n \boldsymbol{F}_m \boldsymbol{W}_{k'} \boldsymbol{F}_n{}^H \boldsymbol{H}_k\Big) + \sum_{u=1}^{U} p_u |e_{u,k}|^2 + \sigma_k^2\Big). \tag{2.8}$$

2.2.2 Uplink Throughput

In uplink, define self-interference channel matrix $\boldsymbol{h}_{SI} \in \mathbb{C}^{L \times L}$, whose element $(\boldsymbol{h}_{SI})_{ij}$ represents the channel from the j-th DA to the i-th DA. Specifically, the diagonal element $(\boldsymbol{h}_{SI})_{ii}$, for $\forall i \in L$, denotes the i-th DA's self-interference channel. The self-interference is the downlink transmitted signal from the DAs, and thus can be calculated as $\sum_{k=1}^{K} \sum_{l=1}^{L} t_l \boldsymbol{F}_l \boldsymbol{w}_k d_k^{DL}$. Define an SIC amount α as the ratio of post-SIC self-interference power over the pre-SIC self-interference power. Thus, the residual self-interference after SIC is calculated as $\frac{1}{\sqrt{\alpha}} \boldsymbol{h}_{SI} \sum_{k=1}^{K} \sum_{l=1}^{L} t_l \boldsymbol{F}_l \boldsymbol{w}_k d_k^{DL}$, and the uplink received signal at the DAs is calculated by

$$\boldsymbol{y}^{UL} = \sum_{u=1}^{U} \sqrt{p_u} \boldsymbol{g}_u d_u^{UL} + \underbrace{\frac{1}{\sqrt{\alpha}} \boldsymbol{h}_{SI} \sum_{k=1}^{K} \sum_{l=1}^{L} t_l \boldsymbol{F}_l \boldsymbol{w}_k d_k^{DL}}_{\text{residual self-interference}} + \boldsymbol{z}, \tag{2.9}$$

where vector $\boldsymbol{g}_u \in \mathbb{C}^{L \times 1}$ is the uplink channel from user u to L DAs. In particular, the element $(\boldsymbol{g}_u)_l$ denotes the channel between the l-th DA and uplink user u. Vector $\boldsymbol{z} \in \mathbb{C}^{L \times 1}, \sim CN(0, \sigma^2 \boldsymbol{I}_L)$ denotes the AWGN receive noise. Assume that the central unit employs a linear maximum ratio combining (MRC) receiver [38] $\boldsymbol{v}_u \in \mathbb{C}^{L \times 1}$, for decoding of the u-th uplink user's data, which is calculated as $\boldsymbol{v}_u = \sum_{i=1}^{L} r_i \boldsymbol{F}_i \boldsymbol{g}_u$. It is because MRC can provide a performance close to minimum mean square error (MMSE) receive-beamforming with a large number of antennas [41]. Besides, since the coupling between $\boldsymbol{r}$, $\boldsymbol{t}$, $\boldsymbol{W}_k$ and the residual self-interference at DAs, it is difficult to derive a tractable solution by applying MMSE or zero-forcing receive-beamforming. Since $r_i = 0$ means the receive chain of the i-th DA is deactivated, the equivalent SINR for uplink user u is calculated as (2.10)

$$\Gamma_u^{UL} = \frac{p_u |\boldsymbol{v}_u{}^H \boldsymbol{g}_u|^2}{\sum_{u' \neq u}^{U} p_{u'} |\boldsymbol{v}_u{}^H \boldsymbol{g}_{u'}|^2 + \frac{1}{\alpha} \left| \boldsymbol{v}_u{}^H \boldsymbol{h}_{SI} \sum_{k=1}^{K} \sum_{l=1}^{L} t_l \boldsymbol{F}_l \boldsymbol{w}_k \right|^2 + ||\boldsymbol{v}_u||^2 \sigma^2}. \tag{2.10}$$

Therefore, the overall uplink throughput is calculated as

$$T_{UL} = \sum_{u=1}^{U} \log_2(1 + \Gamma_u^{UL}). \tag{2.11}$$

Define $\boldsymbol{G_u} = \boldsymbol{g_u}\boldsymbol{g_u}^H$ and substitute $\boldsymbol{v_u} = \sum_{i=1}^{L} r_i \boldsymbol{F_i}\boldsymbol{g_u}$ into (2.11), and then the overall uplink throughput is calculated as (2.12)

$$\begin{aligned}
T_{UL} = &\sum_{u=1}^{U} \log_2\Big(\sum_{u'=1}^{U} p_{u'} \mathrm{Tr}(\boldsymbol{G_{u'}} \sum_{i=1}^{L}\sum_{j=1}^{L} r_i r_j \boldsymbol{F_i} \boldsymbol{G_u} \boldsymbol{F_j}^H) \\
&+ \sigma^2 \mathrm{Tr}(\sum_{i=1}^{L}\sum_{j=1}^{L} r_i r_j \mathbf{F}_i \mathbf{G}_u \mathbf{F}_j^H) \\
&+ \frac{1}{\alpha} \mathrm{Tr}\big(\mathrm{diag}(\boldsymbol{h_{SI}} (\sum_{k=1}^{K}\sum_{m=1}^{L}\sum_{n=1}^{L} t_m t_n \boldsymbol{F_m} \boldsymbol{W_k} \boldsymbol{F_n}^H) \boldsymbol{h_{SI}}^H) \\
&(\sum_{i=1}^{L}\sum_{j=1}^{L} r_i r_j \boldsymbol{F_i} \boldsymbol{G_u} \boldsymbol{F_j}^H))\Big) \\
&- \sum_{u=1}^{U} \log_2\Big(\sum_{u' \neq u}^{U} p_{u'} \mathrm{Tr}(\boldsymbol{G_{u'}} \sum_{i=1}^{L}\sum_{j=1}^{L} r_i r_j \boldsymbol{F_i} \boldsymbol{G_u} \boldsymbol{F_j}^H) \\
&+ \sigma^2 \mathrm{Tr}(\sum_{i=1}^{L}\sum_{j=1}^{L} r_i r_j \boldsymbol{F_i} \boldsymbol{G_u} \boldsymbol{F_j}^H) \\
&+ \frac{1}{\alpha} \mathrm{Tr}\big(\mathrm{diag}(\boldsymbol{h_{SI}} (\sum_{k=1}^{K}\sum_{m=1}^{L}\sum_{n=1}^{L} t_m t_n \boldsymbol{F_m} \boldsymbol{W_k} \boldsymbol{F_n}^H) \boldsymbol{h_{SI}}^H) \\
&(\sum_{i=1}^{L}\sum_{j=1}^{L} r_i r_j \boldsymbol{F_i} \boldsymbol{G_u} \boldsymbol{F_j}^H))\Big)
\end{aligned} \tag{2.12}$$

2.2.3 Power Consumption

The power consumption model mainly include PA power, circuit power, system fixed power and SIC power [26, 37, 42].

PA power: The PA power is determined by the drain efficiency of PAs and beamforming design, calculated as $\frac{1}{\omega}\big(\mathrm{Tr}(\sum_{k=1}^{K} \boldsymbol{W_k}) + \sum_{u=1}^{U} p_u\big)$ [38]. ω is the drain efficiency of PAs of BS and multiple users. Without loss of generality, it is assumed that the drain efficiency of all PAs are equivalent.

Circuit power: Circuit power is proportional to the number of activated transmit- and receive-chains, which include the power consumed by digital/analogue converter, filter, synthesiser, electrical/optical converter, etc. Circuit power consumption can be calculated as $\sum_{l=1}^{L}\left((1-r_l)(1-t_l)p_{idle}\right)+\sum_{l=1}^{L}(r_l p_{c,r})+\sum_{l=1}^{L}(t_l p_{c,t})$, where $p_{c,t}$ and $p_{r,t}$ denote the circuit power consumed by active receive- and transmit-chains, respectively. p_{idle} denotes the circuit power consumption of each sleeping DA.

System fixed power: System fixed power consumption p_{fix} is the power incurred by power supply, active cooling system, etc., which is independent of the state of transmit- and receive-chains. In general, the system fixed power consumption is formulated as a constant.

SIC power: For the DAs working in FD mode, additional power is incurred for SIC operation. Write p_{can} as the power required by SIC at each DA working at FD mode, and thus the total power incurred by SIC is calculated as $\sum_{l=1}^{L}(r_l t_l p_{can})$. It is worthy noting that for distributed MIMO, AC and DC of SIC operation are only needed for each FD DA. It is because the self-interference across different DAs can be managed in the propagation domain by beamforming design. As a result, the overall power consumption is calculated as

$$\begin{aligned}P_{\text{total}} &= \underbrace{\frac{1}{\omega}(\mathrm{Tr}(\sum_{k=1}^{K}\boldsymbol{W}_k)+\sum_{u=1}^{U}p_u)}_{\text{PA power}}+\underbrace{\sum_{l=1}^{L}(t_l r_l p_{can})}_{\text{SIC power}}+p_{fix}+\\ &\underbrace{\sum_{l=1}^{L}\left((1-r_l)(1-t_l)p_{idle}\right)+\sum_{l=1}^{L}(r_l p_{c,r})+\sum_{l=1}^{L}(t_l p_{c,t})}_{\text{circuit power}}.\end{aligned} \tag{2.13}$$

2.3 Activation/Deactivation of DAs, Downlink Beamformer and Uplink Transmission Power Design

Based on the throughput and power consumption analysis, the optimization problem $P2.1$ is re-formulated as

$$\begin{aligned}P2.2: \quad &\underset{\boldsymbol{r},\boldsymbol{t},\boldsymbol{W}_k,p_u,k\in K,u\in U,l\in L}{\operatorname{argmax}} \quad \frac{T_{DL}(\boldsymbol{r},\boldsymbol{t},\boldsymbol{W}_k,p_u)+T_{UL}(\boldsymbol{r},\boldsymbol{t},\boldsymbol{W}_k,p_u)}{P_{\text{total}}(\boldsymbol{r},\boldsymbol{t},\boldsymbol{W}_k,p_u)},\\ &s.t.\ (\widetilde{C1}): \sum_{k=1}^{K}\mathrm{Tr}(\boldsymbol{W}_k\boldsymbol{F}_l)\le p_{DA},\forall l\in L,\\ &(C2),(C3),(C4),\\ &(C5): \boldsymbol{W}_k\succeq 0,\forall k\in K,\\ &(C6): \mathrm{Rank}(\boldsymbol{W}_k)=1,\forall k\in K,\end{aligned} \tag{2.14}$$

where constraint ($\widetilde{C1}$) denotes the transmission power at each DA is upper bounded by p_{DA}, for $\forall l \in L$. ($C5$) and ($C6$) are imposed to guarantee that $\boldsymbol{W}_k = \boldsymbol{w}_k \boldsymbol{w}_k^H$ holds.

The problem in $P2.2$ is generally very difficult to solve. It involves (a) vectors $\boldsymbol{r}, \boldsymbol{t}$ with binary elements in the objective function. Furthermore, the combinational constraints in ($C3$), ($C4$) increase the difficulty in finding the optimal solution. (b) product of binary variables t_m, t_n, r_i, r_j and continuous variables $\boldsymbol{W}_k$, p_u, for $\forall k \in K, u \in U$ and $\forall m, n, i, j \in L$. (c) rank one constraint in ($C6$). Besides, with L DAs, there are 4^L possibilities of DA configurations. It is prohibitive to find the global optimum in terms of computational complexity and thus a low-complexity sub-optimal algorithm is desirable. In the sequels, a channel-gain-based algorithm is first proposed to perform DAs activation/deactivation, which effectively removes the binary variables in the objective and in the constraints ($C3$) ($C4$), and also the product between binary variables and continuous variables. After the activation/deactivation of DAs, a novel D-HD algorithm is proposed to optimize beamformer $\boldsymbol{W}_k$, for $\forall k \in K$ and transmission power p_u, for $\forall u \in U$. At last, total complexity is given.

2.3.1 *Activation/Deactivation of Transmit/Receive Chains at DAs*

If the channel gain between the l-th DA and uplink users has good quality, while the channel gain between the DA and downlink users is poor, the l-th DA can work in HD mode for reception only. Since if the DA works in FD mode, its downlink contribution is marginal, while additional power consumption is incurred for FD mode. Also, downlink transmission interferes uplink reception due to the introduced self-interference. Similarly, if the channel gain between the DA and uplink users is poor while the channel gain between the DA and downlink users has good quality, the DA can operate in HD for signal transmission only. If one DA has good channel quality in both uplink and downlink, it can work in FD mode.

To implement activation and deactivation for DAs, a threshold, i.e., ψ, is introduced. The threshold can be adjusted according to different QoS requirements and DA density. For example, one can increase the value of ψ if power consumption requirement is stringent. It means less DAs will be activated and therefore power consumption is decreased. On the other hand, if users' SE requirement is stringent, one can decrease the value of ψ. Also, for a dense DA deployment, the value of the threshold may be higher than a sparse deployment. The DA activation and deactivation algorithm is summarized in Algorithm 1.

Algorithm 1 DA activation and deactivation algorithm

Input: $\boldsymbol{h}_k, \boldsymbol{g}_u$ for $\forall k \in k, u \in U$, and the threshold ψ.
Output: Transmit/receive chains vector $\boldsymbol{r}^*, \boldsymbol{t}^*$.

for $l = 1, ..., L$ **do**
  for $u = 1, ..., U$ **do**
    Calculate the channel condition between the l-th DA and uplink user u.
    if $(\boldsymbol{g}_u)_l \geq \psi$ **then**
      $r_l = 1$, break.
    else
      $r_l = 0$.
    end if
  end for
  for $k = 1, ..., K$ **do**
    Calculate the channel condition between the l-th DA and downlink user k.
    if $(\boldsymbol{h}_k)_l \geq \psi$ **then**
      $t_l = 1$, break.
    else
      $t_l = 0$.
    end if
  end for
end for
return $\boldsymbol{r}^*, \boldsymbol{t}^*$.

2.3.2 *Design of Downlink Beamformer at DAs and Transmission Power at Uplink Users*

For ease of notation, a matrix $\boldsymbol{W} = \{\boldsymbol{W}_1, \boldsymbol{W}_2, ..., \boldsymbol{W}_K\}$ is introduced to include all beamformers $\boldsymbol{W}_k, \forall k \in K$. Define a vector $\boldsymbol{p} = \{p_1, p_2, ..., p_U\}$ to include transmission power variables of all uplink users $\forall u \in U$. The feasible domain confined by the constraints is denoted as $\boldsymbol{\Theta}$. Then the optimization problem is equivalently written as

$$P2.3 : \underset{\boldsymbol{W}, \boldsymbol{p} \in \boldsymbol{\Theta}}{\text{argmax}} \frac{T_{DL}(\boldsymbol{W}, \boldsymbol{p}) + T_{UL}(\boldsymbol{W}, \boldsymbol{p})}{P_{\text{total}}(\boldsymbol{W}, \boldsymbol{p})}, \qquad (2.15)$$
$$s.t. \ (\widetilde{C1}), (C2), (C5), (C6).$$

For the overall throughput $T_{DL}(\boldsymbol{W}, \boldsymbol{p}) + T_{UL}(\boldsymbol{W}, \boldsymbol{p})$, all the positive parts are collected into $f_1(\boldsymbol{W}, \boldsymbol{p})$, while the negative parts are collected into $f_2(\boldsymbol{W}, \boldsymbol{p})$. Then the overall throughput can be expressed as $f_1(\boldsymbol{W}, \boldsymbol{p}) - f_2(\boldsymbol{W}, \boldsymbol{p})$, where $f_1(\boldsymbol{W}, \boldsymbol{p})$ and $f_2(\boldsymbol{W}, \boldsymbol{p})$ are given by (2.16) and (2.17), respectively.

$$
\begin{aligned}
f_1(\boldsymbol{W}, \boldsymbol{p}) = & \sum_{k=1}^{K} \log_2 \Big(\sum_{k'=1}^{K} \mathrm{Tr}(\sum_{m=1}^{L}\sum_{n=1}^{L} t_m t_n \boldsymbol{F_m} \boldsymbol{W_{k'}} \boldsymbol{F_n}^H \boldsymbol{H_k}) + \sum_{u=1}^{U} p_u |e_{u,k}|^2 + \sigma_k^2 \Big) \\
& + \sum_{u=1}^{U} \log_2 \Big(\sum_{u'=1}^{U} p_{u'} \mathrm{Tr}(\boldsymbol{G_{u'}} \sum_{i=1}^{L}\sum_{j=1}^{L} r_i r_j \boldsymbol{F_i} \boldsymbol{G_u} \boldsymbol{F_j}^H) \\
& + \sigma^2 \mathrm{Tr}(\sum_{i=1}^{L}\sum_{j=1}^{L} r_i r_j \boldsymbol{F_i} \boldsymbol{G_u} \boldsymbol{F_j}^H) \\
& + \frac{1}{\alpha} \mathrm{Tr}\big(\mathrm{diag}(\boldsymbol{h_{SI}} (\sum_{k=1}^{K}\sum_{m=1}^{L}\sum_{n=1}^{L} t_m t_n \boldsymbol{F_m} \boldsymbol{W_k} \boldsymbol{F_n}^H) \boldsymbol{h_{SI}}^H) \\
& (\sum_{i=1}^{L}\sum_{j=1}^{L} r_i r_j \boldsymbol{F_i} \boldsymbol{G_u} \boldsymbol{F_j}^H))\Big).
\end{aligned} \tag{2.16}
$$

$$
\begin{aligned}
f_2(\boldsymbol{W}, \boldsymbol{p}) = & \sum_{k=1}^{K} \log_2 \Big(\sum_{k' \neq k}^{K} \mathrm{Tr}(\sum_{m=1}^{L}\sum_{n=1}^{L} t_m t_n \boldsymbol{F_m} \boldsymbol{W_{k'}} \boldsymbol{F_n}^H \boldsymbol{H_k}) + \sum_{u=1}^{U} p_u |e_{u,k}|^2 + \sigma_k^2 \Big) \\
& + \sum_{u=1}^{U} \log_2 \Big(\sum_{u' \neq u}^{U} p_{u'} \mathrm{Tr}(\boldsymbol{G_{u'}} \sum_{i=1}^{L}\sum_{j=1}^{L} r_i r_j \boldsymbol{F_i} \boldsymbol{G_u} \boldsymbol{F_j}^H) \\
& + \sigma^2 \mathrm{Tr}(\sum_{i=1}^{L}\sum_{j=1}^{L} r_i r_j \boldsymbol{F_i} \boldsymbol{G_u} \boldsymbol{F_j}^H) \\
& + \frac{1}{\alpha} \mathrm{Tr}\big(\mathrm{diag}(\boldsymbol{h_{SI}} (\sum_{k=1}^{K}\sum_{m=1}^{L}\sum_{n=1}^{L} t_m t_n \boldsymbol{F_m} \boldsymbol{W_k} \boldsymbol{F_n}^H) \boldsymbol{h_{SI}}^H) \\
& (\sum_{i=1}^{L}\sum_{j=1}^{L} r_i r_j \boldsymbol{F_i} \boldsymbol{G_u} \boldsymbol{F_j}^H))\Big).
\end{aligned} \tag{2.17}
$$

$f_1(\boldsymbol{W}, \boldsymbol{p})$ and $f_2(\boldsymbol{W}, \boldsymbol{p})$ are both jointly-concave with respect to the variables $\boldsymbol{W}, \boldsymbol{p}$ in the feasible domain $\boldsymbol{\Theta}$. Hence, the FW method can be used to handle the difference of two concave functions. The FW method approximates of the minus part $f_2(\boldsymbol{W}, \boldsymbol{p})$ by its first order Taylor series $f_2^{(n)}(\boldsymbol{W}, \boldsymbol{p})$, and updates the approximation $f_2^{(n)}(\boldsymbol{W}, \boldsymbol{p})$ iteratively along the direction that approaches the original function $f_2(\boldsymbol{W}, \boldsymbol{p})$ [24]. Suppose that the value of $\boldsymbol{W}, \boldsymbol{p}$ is denoted by $\boldsymbol{W}^{(n)}, \boldsymbol{p}^{(n)}$, and the value of $f_2(\boldsymbol{W}, \boldsymbol{p})$ is approximated by $f_2^{(n)}(\boldsymbol{W}, \boldsymbol{p})$ at the n-th iteration. Since the total throughput $f_1(\boldsymbol{W}, \boldsymbol{p}) - f_2^{(n)}(\boldsymbol{W}, \boldsymbol{p})$ is the lower bound of the original value $f_1(\boldsymbol{W}, \boldsymbol{p}) - f_2(\boldsymbol{W}, \boldsymbol{p})$, the variables $\boldsymbol{W}^{(n)}, \boldsymbol{p}^{(n)}$ are iteratively updated and the lower bound increases after every iteration. Because of the power consumption is upper bounded by constraints, the iterative procedure can guarantee the convergence. Since $f_2(\boldsymbol{W}, \boldsymbol{p})$ is concave and differentiable on the considered domain, one can easily find its first-order approximation by (2.18), such that

$$
\begin{aligned}
f_2^{(n)}(\boldsymbol{W}, \boldsymbol{p}) &= f_2(\boldsymbol{W}^{(n)}, \boldsymbol{p}^{(n)}) + \sum_{k=1}^{K}(a_k^{(n)})^{-1}\sum_{u=1}^{U}(p_u - p_u^{(n)})|e_{u,k}|^2 \\
&+ \sum_{k=1}^{K}(a_k^{(n)})^{-1}\sum_{k'\neq k}\mathrm{Tr}\Big(\sum_{m=1}^{L}\sum_{n=1}^{L} t_m t_n \boldsymbol{F_m}(\boldsymbol{W_{k'}} - \boldsymbol{W_{k'}^{(n)}})\boldsymbol{F_n}^H \boldsymbol{H_k}\Big) \\
&+ \sum_{u=1}^{U}(b_u^{(n)})^{-1}\sum_{u'\neq u}(p_{u'} - p_{u'}^{(n)})\mathrm{Tr}\Big(\boldsymbol{G_{u'}}\sum_{i=1}^{L}\sum_{j=1}^{L} r_i r_j \boldsymbol{F_i}\boldsymbol{G_u}\boldsymbol{F_j^H}\Big) \\
&+ \sum_{u=1}^{U}\Bigg((b_u^{(n)})^{-1}\frac{1}{\alpha}\mathrm{Tr}\Big(\mathrm{diag}\big(\boldsymbol{h_{SI}}\big(\sum_{k=1}^{K}\sum_{m=1}^{L}\sum_{n=1}^{L} t_m t_n \boldsymbol{F_m}(\boldsymbol{W_k} - \boldsymbol{W_k}^{(n)})\boldsymbol{F_n}^H\big)\boldsymbol{h_{SI}}^H\big) \\
&\quad \big(\sum_{i=1}^{L}\sum_{j=1}^{L} r_i r_j \boldsymbol{F_i}\boldsymbol{G_u}\boldsymbol{F_j}^H\big)\Big)\Bigg)
\end{aligned} \tag{2.18}
$$

where we have

$$
(a_k^{(n)}) = \sum_{k'\neq k}\mathrm{Tr}\Big(\sum_{i=1}^{L}\sum_{j=1}^{L} t_m t_n \boldsymbol{F_m}\boldsymbol{W_{k'}}^{(n)}\boldsymbol{F_n}^H \boldsymbol{H_k}\Big) + \sum_{u=1}^{U} p_u^{(n)}|e_{u,k}|^2 + \sigma_k^2, \tag{2.19}
$$

$$
\begin{aligned}
(b_u^{(n)}) &= \sigma^2\mathrm{Tr}\Big(\sum_{i=1}^{L}\sum_{j=1}^{L} r_i r_j \boldsymbol{F_i}\boldsymbol{G_u}\boldsymbol{F_j}^H\Big) + \sum_{u'\neq u}(p_{u'}^{(n)})\mathrm{Tr}\Big(\sum_{i=1}^{L}\sum_{j=1}^{L} r_i r_j \boldsymbol{F_i}\boldsymbol{G_u}\boldsymbol{F_j}^H\Big) + \\
&\frac{1}{\alpha}\mathrm{Tr}\Big(\mathrm{diag}\big(\boldsymbol{h_{SI}}\big(\sum_{k=1}^{K}\sum_{m=1}^{L}\sum_{n=1}^{L} t_m t_n \boldsymbol{F_m}\boldsymbol{W_k}^{(n)}\boldsymbol{F_n}^H\big)\boldsymbol{h_{SI}}^H\big)\big(\sum_{i=1}^{L}\sum_{j=1}^{L} r_i r_j \boldsymbol{F_i}\boldsymbol{G_u}\boldsymbol{F_j}^H\big)\Big).
\end{aligned} \tag{2.20}
$$

Now, the overall throughput is re-expressed as $f_1(\boldsymbol{W}, \boldsymbol{p}) - f_2^{(n)}(\boldsymbol{W}, \boldsymbol{p})$, which is jointly-concave with respect to the variables. It is because $f_1(\boldsymbol{W}, \boldsymbol{p})$ is jointly-concave, and $f_2^{(n)}(\boldsymbol{W}, \boldsymbol{p})$ is affine with respect to all variables $\boldsymbol{W}, \boldsymbol{p} \in \{\boldsymbol{\Theta}\}$. The optimization problem can be expressed as

$$
\begin{aligned}
P2.4 : & \underset{\boldsymbol{W}, \boldsymbol{p} \in \boldsymbol{\Theta}.}{\text{argmax}} \frac{f_1(\boldsymbol{W}, \boldsymbol{p}) - f_2^{(n)}(\boldsymbol{W}, \boldsymbol{p})}{P_{\text{total}}(\boldsymbol{W}, \boldsymbol{p})}, \\
& s.t. \ (\widetilde{C1}), (C2), (C5), (C6).
\end{aligned}
\tag{2.21}
$$

Since the problem P2.4 is calculated as the ratio between a concave function and an affine function in the consider domain, Theorem 2.1 is introduced to solve the problem.

Theorem 2.1 *The reformulated EE,* $\frac{f_1(\boldsymbol{W}, \boldsymbol{p}) - f_2^{(n)}(\boldsymbol{W}, \boldsymbol{p})}{P_{\text{total}}(\boldsymbol{W}, \boldsymbol{p})}$, *is jointly quasi-concave with respect to the variables* $\boldsymbol{W}, \boldsymbol{p}$ *in the feasible domain.*

Proof of Theorem 2.1 is given in Appendix A.

Theorem 2.1 confirms that there is a global optimal EE in the feasible domain. For the fractional structured quasi-concave problem, $\beta = \frac{f_1(\boldsymbol{W},\boldsymbol{p}) - f_2^{(n)}(\boldsymbol{W},\boldsymbol{p})}{P_{\text{total}}(\boldsymbol{W},\boldsymbol{p})}$, it can be associated with a subtract programming [43, 44], such that

$$
f_1(\boldsymbol{W}, \boldsymbol{p}) - f_2^{(n)}(\boldsymbol{W}, \boldsymbol{p}) - \beta p_{\text{total}}(\boldsymbol{W}, \boldsymbol{p}) \tag{2.22}
$$

Assume that β^* is the optimal value of (2.22), it is obvious that maximizing (2.22) is equivalent to finding the root of $f_1(\boldsymbol{W}, \boldsymbol{p}) - f_2^{(n)}(\boldsymbol{W}, \boldsymbol{p}) - \beta^* P_{\text{total}}(\boldsymbol{W}, \boldsymbol{p})$, as proven in Appendix B. Therefore, with the equivalent subtract programming, the problem becomes into solving $f_1(\boldsymbol{W}, \boldsymbol{p}) - f_2^{(n)}(\boldsymbol{W}, \boldsymbol{p}) - \beta^* P_{\text{total}}(\boldsymbol{W}, \boldsymbol{p})$ with a given β.

$$
\begin{aligned}
P2.5 : & \underset{\boldsymbol{W}, \boldsymbol{p} \in \boldsymbol{\Theta}.}{\text{argmax}} f_1(\boldsymbol{W}, \boldsymbol{p}) - f_2^{(n)}(\boldsymbol{W}, \boldsymbol{p}) - \beta P_{\text{total}}(\boldsymbol{W}, \boldsymbol{p}), \\
& s.t. \ (\widetilde{C1}), (C2), (C5), (C6).
\end{aligned}
\tag{2.23}
$$

Rank relaxation can be used to handle the rank constraint [43, 45]. To be specific, when the channels are statistically independent, the solution of original problem is rank-one. The proof is demonstrated in Appendix C. In DA systems, as the antennas are geographically distributed, the channel is more likely to be independent [37]. In other words, after obtaining the optimal matrices $\boldsymbol{W}$, it has a high probability satisfying the rank one constraint, and thus can be decoupled into beamformers.

After rank one relaxation, now P2.5 is a standard semi-definite programming (SDP) problem. Therefore, the CVX solver can be used to obtain the optimal result. In a nutshell, the algorithm solves the SDP problem P2.5 in the inner layer and updates β in the outer layer (The convergence in outer and inner layers is proven in Appendices D and E). The D-HD algorithm is summarized as Algorithm 2.

Algorithm 2 D-HD algorithm

Input: Antenna configuration vectors $\boldsymbol{r}, \boldsymbol{t}$, left/right bounds β_l and β_r, channel condition, i.e., $\boldsymbol{h_k}, g_u, e_{u,k}, \boldsymbol{h_{SI}}$, for $\forall k \in k, u \in U$, and power consumption parameters, i.e., $\eta, p_{idle}, p_{c,t}, p_{c,r}, p_{can}, p_{fix}, p_{DA}, p_{u,max}$.

Output: Optimal beamforming weight $\boldsymbol{W}_k^*$, for $\forall k \in k$, and optimal uplink transmission power $p_u^*, \forall u \in U$.

1: Set accuracy factor $\epsilon > 0$, and suppose $\mathcal{F}(\beta)$ is the optimal value of $f_1(\boldsymbol{W}, \boldsymbol{p}) + f_2^{(n)}(\boldsymbol{W}, \boldsymbol{p}) - \beta P_{\text{total}}(\boldsymbol{W}, \boldsymbol{p})$. Initialize left bound β_l and right bound β_r that ensure the $\mathcal{F}(\beta_l) \cdot \mathcal{F}(\beta_r) < 0$.
2: **while** $\beta_r - \beta_l > \epsilon$ **do**
3: $\beta = \frac{\beta_r + \beta_l}{2}$.
4: Solve the problem $P5$ with the FW method until convergence.
5: **if** $\mathcal{F}(\beta_l) \cdot \mathcal{F}(\beta) < 0$ **then**
6: $\beta_r = \beta$.
7: **else**
8: $\beta_l = \beta$.
9: **end if**
10: **end while**

2.4 Complexity Analysis

In the process of activation of deactivation of DAs, at most $N(K + U)$ comparisons is required to find $\boldsymbol{r}^*$ and $\boldsymbol{t}^*$. After this, the D-HD algorithm is performed to design beamformer at DAs and transmission power at uplink users. The fractional-structure $\frac{f_1(\boldsymbol{W}, \boldsymbol{p}) - f_2^{(n)}(\boldsymbol{W}, \boldsymbol{p})}{P_{\text{total}}(\boldsymbol{W}, \boldsymbol{p})}$ is transformed into a difference-structure $f_1(\boldsymbol{W}, \boldsymbol{p}) - f_2^{(n)}(\boldsymbol{W}, \boldsymbol{p}) - \beta P_{\text{total}}(\boldsymbol{W}, \boldsymbol{p})$. Bisection search can be used in the outer layer for finding optimal β^*. With ϵ as the accuracy factor, at most $\log_2(\frac{\beta_r - \beta^*}{\epsilon})$ iterations can guarantee convergence. In the inner layer, CVX is used to solve the SDP problem $f_1(\boldsymbol{W}, \boldsymbol{p}) - f_2^{(n)}(\boldsymbol{W}, \boldsymbol{p}) - \beta P_{\text{total}}(\boldsymbol{W}, \boldsymbol{p})$, where complexity is indexed as ξ. Since the FW method is adopted for the inner layer iteration, the convergence rate is $\mathcal{O}(\frac{1}{n})$ [46]. In the n-th update in the inner layer, the CVX solver uses an interior-point method to solve the SDP problem, which belongs to the class of path-following method and leads a fast convergence. Hence, the complexity of the D-HD algorithm is calculated $\mathcal{O}(\log_2(\frac{\beta_r - \beta^*}{\epsilon}) \cdot \xi)$. As a result, the overall complexity is calculated by $\mathcal{O}\big(N(K + U) + \log_2(\frac{\beta_r - \beta^*}{\epsilon}) \cdot \xi\big)$, where the bisection search in the outer layer has the largest impact on the complexity of the whole algorithm.

2.5 Performance Comparison

The simulation setup is shown in Table 2.2. The 3GPP PL model PL $= 145.4 + 37.5\log_{10}(d/1000)$ is adopted [47]. There are two comparison algorithms. The first one is the distributed FD (D-FD) [38], where all DAs can only work in FD or sleep modes. The other is co-located FD (C-FD) MIMO [43]. In particular, all

Table 2.2 Simulation setup of bi-directional FD DA systems

Central carrier frequency	2 GHz
Bandwidth	1 MHz
AWGN power spectral density	−174 dBm/Hz
Cell model	Square grid 100 m^2
DE of PA η	25%
Rician factor	5 dB
p_{idle}, $p_{c,r}$ and $p_{c,t}$	10, 100, and 100 mW
p_{can} and p_{fix}	50 and 500 mW
p_{DA} and $p_{u,max}$	100 and 40 mW
Self-interference cancellation amount α	$\alpha = 60$–100 dB
Antenna gain	0 dBi
Number of users	4
Uplink users' location	(92,10), (85,86)
Downlink users' location	(5,76), (78,16)

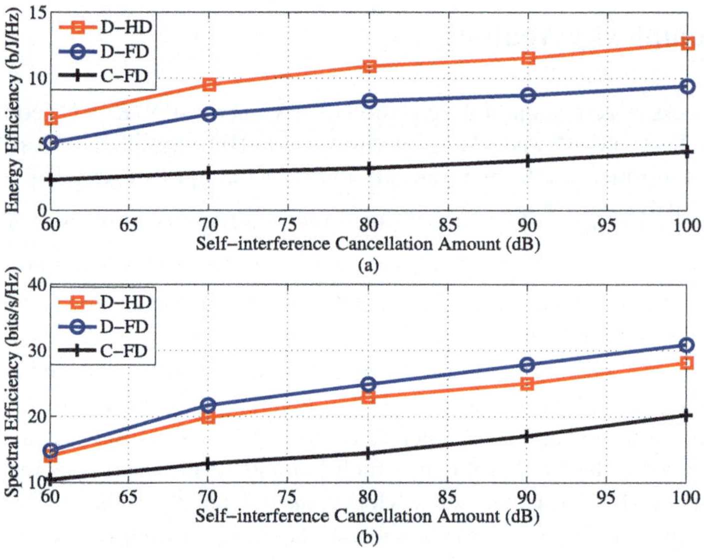

Fig. 2.3 Average EE and SE performance vs. different SIC amount, with 16 DAs and $\psi = 0.5 \times 10^{-8}$

antennas have identical PL to users due to the co-located antenna deployment, and the activation (or deactivation) of transmit/receive-chains is disabled.

Figure 2.3a demonstrates the average optimal EE performance under D-HD, D-FD and C-FD systems. The centralized antennas of C-FD system are located in the center of the considered area. It is demonstrated that the designed D-HD algorithm outperforms others in terms of EE at all SIC regions. It is because under the proposed D-HD algorithm, the DAs with poor channel condition are deactivated, to avoid unnecessary power consumption.

Besides, the activated DAs can work in HD modes (only transmit or receive) if the DAs have only reasonable channel quality in one direction. Differently, under the D-FD algorithm, all activated DAs work in FD mode. This leads to much higher power consumption than the proposed D-HD, but not all DAs can contribute reasonable throughput in bi-directions, hence the EE is degraded. For the C-FD with centralized antennas, all co-located antennas have similar PL, leading to poor throughput performance once users are located far from the antennas. Besides, high power consumption is incurred due to the fully activated antennas. As a result, the EE performance is the lowest among the three systems. Also, it is investigated that, for all the three systems, EE can be improved with increased SIC amount, as a lower level of residual self-interference provides better communication performance. Figure 2.3b shows the average SE for the three systems. D-FD shows the highest SE performance since all antennas work in FD. With a reasonable level of SIC, its SE performance be higher than others. However, the designed D-HD only shows marginal SE loss compared to the D-FD, but incurs much lower power consumption. The C-FD algorithm demonstrates the lowest SE performance. It is because the co-located antenna structure is sensitive to the distance between users and antennas. Also, due to the poor performance of SIC in propagation domain, the co-located antenna deployment has also resulted in the strongest self-interference among different antennas. As a result, the C-FD SE increases slowly than the DA structure.

Figure 2.4 shows the required number of self-interference reference chains by the three systems. Since the analogue and digital domain cancellation needs to tap reference signal from every transmit chain and feed it to all receive chains for SIC operation [3], the required number of self-interference reference chains has a significant impact on the SIC design. It is demonstrated that, the hybrid duplexing DA system requires the fewest active self-interference reference chains. It is because

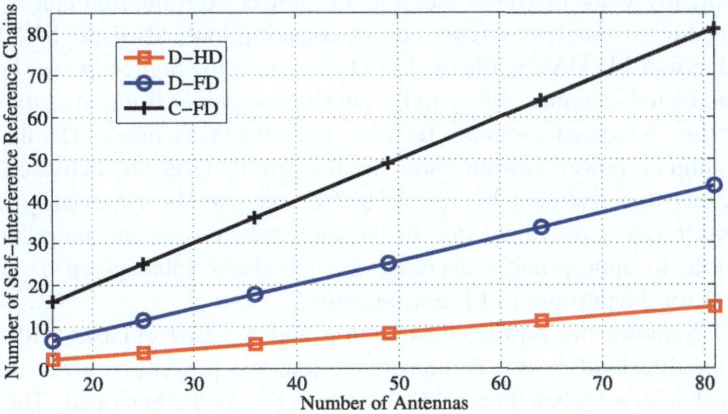

Fig. 2.4 Average required number of reference chains for SIC operation, with $\psi = 1 \times 10^{-9}$

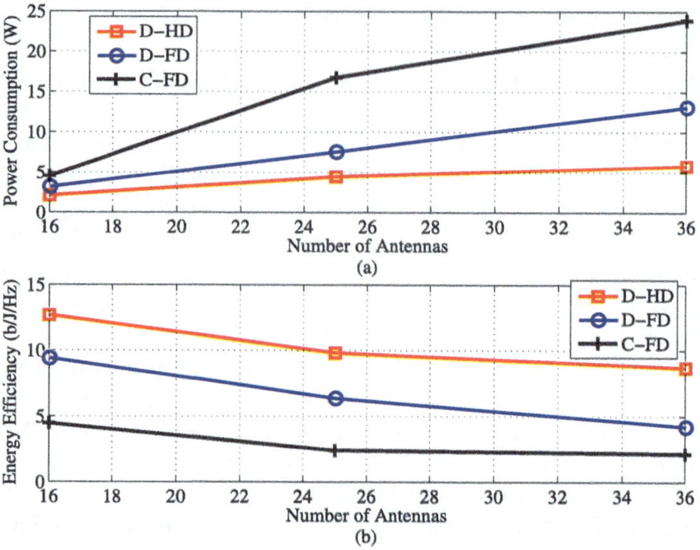

Fig. 2.5 Average power consumption and EE vs. different number of antennas, with $\alpha = 100$ dB and $\psi = 1 \times 10^{-9}$

DAs may sleep or work in HD mode, where SIC operation is not required. Besides, as a advantage of distributed structure, propagation domain cancellation guarantees a reasonable SIC amount across different DAs, and thus the analogue and digital domain SIC may not be required across different DAs. By contrast, all antennas are active and work in FD mode by the conventional C-FD system, where the number of active self-interference reference chains increases linearly with numbers of antennas.

Figure 2.5a demonstrates the overall power consumption under different number of antennas. The proposed D-HD consumes the lowest power. Since the DAs can dynamically work in HD or sleep mode, power consumption can be reduced effectively due to the less circuit power consumption. Also, no SIC power is consumed in non-FD DAs, while the C-FD consumes the highest power because of the fully-activated antennas. Also, high transmission power is required to overcome the PL by the co-located structure. Besides, it shows that a denser DA deployment results in higher power consumption because more DAs are activated. On the other side, as demonstrated by Fig. 2.5b, EE may not be necessarily improved with a denser DA deployment due to the increased power consumption. Hence, it is possible to appropriately decrease the threshold value given a denser DA deployment for the purpose of EE optimization.

Figure 2.6 shows the power consumption and SE performance with different values of the threshold ψ. It is demonstrated that, less power and lower throughput are obtained with a higher threshold, since fewer DAs are activated. This denotes that the threshold value can be adjusted to reduce the power consumption, at the cost

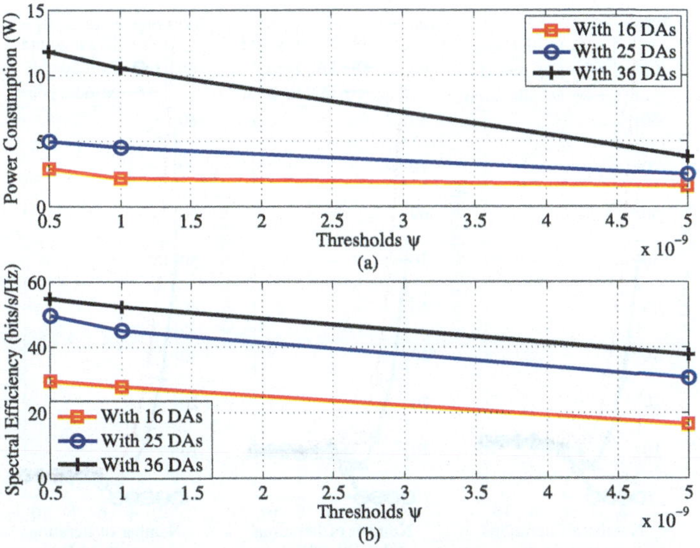

Fig. 2.6 Average power consumption and SE performance vs. different DAC threshold values, with $\alpha = 100$ dB

of SE. On the other hand, a lower threshold allows a better SE performance, while power consumption is relatively higher, which corresponds to a stringent SE QoS scenario. This confirms that by adjusting the value of the threshold ψ, the number of activated DAs can be well controlled towards different QoS requirement. Besides, with more DAs, SE can be significantly improved, while power consumption is increased. As a result, the EE may not be always necessarily improved, which shows the similar observation in Fig. 2.5b.

Figure 2.7 shows the iteration trends of the D-HD algorithm with different number of DAs. It is shown that at most around 10 iterations are required to approach convergence, which confirms the low computational complexity of the proposed algorithm.

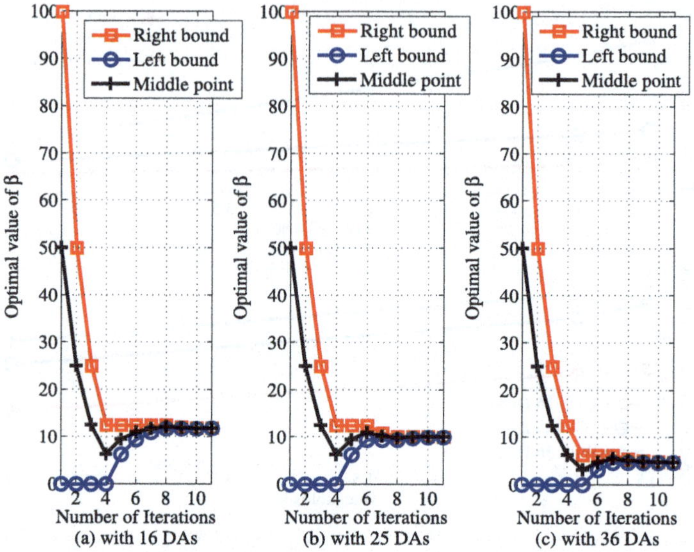

Fig. 2.7 The number of iterations on finding optimal β^* by bisection searching, with left bound $\beta_l = 0$ and right bound $\beta_r = 100$

Chapter 3
Energy Efficient Wireless-Powered Full-Duplex Relaying Communications

Wireless power transform (WPT) is an promising technique in scenarios where battery replacement or charging is inconvenient or even impossible [48, 49]. By harvesting energy from the signal source, WPT relay nodes can effectively prolong lifetime and thus help maintain the network connectivity [50]. For conventional WPT relay systems working at HD mode, three time slots are required for data transmission and energy harvesting [51, 52]. To be specific, the first time slot is used for harvesting energy at the relay. Then, the remaining two time slots are needed for forwarding signal from source to destination, as that of typical two-hop HD relay systems. Based on the power splitting energy harvesting structure [53], another WPT HD relay system only needs two time slots. In the first time slot, the source transmits signal to the relay, where a part of the received signal is decoded and forwarded to the destination, while the rest of the received signal is harvested by the relay. Outage occurs when the received signal-to-interference-and-noise-ratio (SINR) is lower than a required threshold [52, 54]. The outage probability in WPT cognitive radio systems was considered in [55], where the cognitive transmitter communicates with the cognitive receiver via an energy harvesting relay. The authors in [56] investigated a three-node system, where a HD relay was deployed to assist communication between source and destination. Also, the power splitting factor at the relay node was optimized to minimize outage probability. In [57], WPT was investigated for OFDMA multiuser systems, where the downlink users split part of the received signal for energy harvesting while the remaining signal is used for signal decoding. In a nutshell, the conventional WPT HD relaying either needs three time slots [51, 52] or forwards a part of signal to the destination [53], incurring SE loss due to the low resource utilization efficiency.

FD can be employed to improve system performance of WPT relaying [53, 58–60]. In [58] and [60], energy harvesting and information transmission are decoupled into two orthogonal time slots. Since the self-interference is treated as noise by the FD relay, SIC is needed at the FD relay. In [59], joint design of source and relay beamforming was investigated in FD MIMO relay systems. The authors in

Z. Wei et al., *Energy Efficient Full Duplex Wireless Communication Systems*,
Wireless Networks, https://doi.org/10.1007/978-981-97-8915-3_3

[61] and [62] researched the resource allocation in bi-directional WPT-aided FD systems, where a FD BS sends the energy to multiple users while the users send information to the FD BS concurrently. Since the self-interference is introduced at the BS, SIC is needed at the BS to guarantee the uplink receive-quality. EE issue in bi-directional WPT FD systems was investigated in [63]. In the first time slot, a BS sends information and energy signal to multiple user at downlink. In the second time slot, the BS continues sending information bearing signal to the downlink user and simultaneously receive information signal from the uplink user. Hence, the BS in fact works in HD during the second time slot. The review of WPT-aided FD/HD relay systems is summarized in Table 3.1.

In conventional WPT FD relay systems, self-interference is treated as a harmful element, which should be suppressed as much as possible. In a different manner, [73, 76] proposed to utilize self-interference as a green source for energy harvesting. The relay receives signal in the first time slot, and at the same time forwards signal to the destination. Since self-interference is introduced to the relay's receiver, it can harvest energy from the self-interference as well as signal from the source in the second time slot. Nevertheless, uniform time slot duration is assumed for the two time slots in [73, 76], which ignores the fact that the received SINR of the source-relay and the relay-users links can hardly be equivalent in WPT relay systems. It is because the relay's transmission power could be much lower than that at the source, due to the wireless channel attenuation from the source to the relay and the energy harvesting efficiency at the relay node. As a result, the end-to-end SE is bounded by the relay-users link while a large proportion of the SE in the source-relay link is wasted in the symmetric transmission system. Besides, the system models proposed in [58–60, 73, 76] only consider a signal-user and single-carrier scenario, which is not applicable to a multiuser scenario, as illustrated in Fig. 3.1.

3.1 System Model for WPT FD Relay Multiuser Systems

3.1.1 System Model

Assume that there are K users and N subcarriers, where OFDMA is used for multiuser access [78]. The channel condition is known at the source and the relay node [64]. The relay has no power supplement, and therefore it needs to harvest energy through WPT from the source [51, 58]. In the first time slot T_1, the source sends information signal to the relay. In the second time slot of duration T_2, the received signal at the relay is decoded, re-encoded and forwarded to the user, and wireless energy is sent from the source to the relay's receiver for energy harvesting concurrently. Write power allocation matrices at the source $\boldsymbol{p_s} = [p_{s,k,n}]_{K\times N}$ and at the relay $\boldsymbol{p_r} = [p_{r,k,n}]_{K\times N}$, where elements $p_{s,k,n}$ and $p_{r,k,n}$ denote the transmission power allocated at the resource and the relay node for user k on subcarrier n, respectively. Write subcarrier allocation matrix $\boldsymbol{\beta} = [\beta_{k,n}]_{K\times N}$, whose

Table 3.1 Summary of WPT-aided FD/HD relay systems

References	Duplex mode	System model	Splitting structure	Optimization metric
[64]	HD	Point-to-point MIMO WPT system at downlink	Time splitting	Harvested power
[51]	HD	Three-node model with a WPT-aided AF relay	Power/time splitting	SE
[65]	HD	Three-node model with a WPT-aided AF relay	Time splitting	SE
[66]	HD	Three-node model. Both the user and relay are WPT-aided	Time/power splitting	SE
[67]	HD	Multi-relay based two-hop systems Relay selection is involved and SI is mitigated as noise	Power splitting	SE
[52]	HD	One source-destination pair and multiple randomly deployed AF relays	power splitting	Outage probability
[68]	HD	Time-splitting based uplink transmission Three time slots are required. Uplink for signal transmission and downlink for WPT	Time splitting	SE
[69]	HD	Downlink multiuser system with imperfect SIC	Power splitting	Outage probability Power consumption
[70]	HD	Multi-input single-output (MISO) system with a transmitter, An legitimate receiver, an EH receiver, and an eavesdropper	/	Secrecy rate
[57]	HD	Point-to-multi-point OFDMA based system at downlink	Power splitting	SE

(continued)

Table 3.1 (continued)

References	Duplex mode	System model	Splitting structure	Optimization metric
[71]	HD	Multiuser MISO systems Multiple information receivers and EH receivers at downlink	Power splitting	Harvested power
[53]	HD	Multiple adjacent AF MIMO multi-way relay networks, each having multiple users and one relay.	Time splitting	SE
[50]	HD	MISO AF relay network with one source, one Relay, one legitimate destination, one eavesdropper, and one energy receiver	/	Secrecy rate
[54]	HD	Multi-relay cooperative system	Time splitting	Outage probability
[55]	HD	Cooperative CR network with a primary receiver, a cognitive transmitter-receiver pair and a WPT-aided relay	power splitting	Outage probability
[56]	HD	Three-node model with a WPT-aided AF relay	power splitting	Outage probability
[72]	HD	AWGN channel with a HD transceiver	Time splitting	Delay
[73]	Pseudo FD	Three-node model with a WPT-aided AF relay. Multiple antennas are equipped in source and relay. SI is recycled as energy.	time splitting	SE
[63]	Pseudo FD	Bi-directional FD with a BS, multiple downlink and uplink users. SI is mitigated as noise by the BS	Time splitting	EE/SE

(continued)

Table 3.1 (continued)

References	Duplex mode	System model	Splitting structure	Optimization metric
[74]	Pseudo FD	Three-node model with a WPT-aided relay node SI is mitigated as noise	power splitting	SINR
[75]	Pseudo FD	P-to-P bi-directional FD systems and SI is mitigated as noise.	Power splitting	Weighted SE
[58]	Pseudo FD	Dual-hop FD AF relay system The relay harvests energy in the first time slot. FD mode is realized in the second time slot.	Time splitting	SE
[59]	Pseudo FD	Beamforming involved at the source and AF relay	Power splitting	SE
[60]	Pseudo FD	Two-hop MIMO AF relay system. SI is mitigated as noise by the relay.	time splitting	SE
[76]	Pseudo FD	Three-node models with symmetric time slots SI is recycled by the AF relay node.	Time splitting	SE
[61]	Pseudo FD	Multiple users are powered by WPT. Uplink operates in time-division HD and SI is mitigated as noise.	Time splitting	SE
[77]	/	Bi-directional FD system with one BS, one uplink user, one downlink user, and one idle user with WPT capability.	/	Secrecy rate
[62]	Pseudo FD	Bi-directional FD with a access point (AP) and multiple users Perfect SIC is assumed and SI is mitigated as noise by the AP.	Time splitting	EE

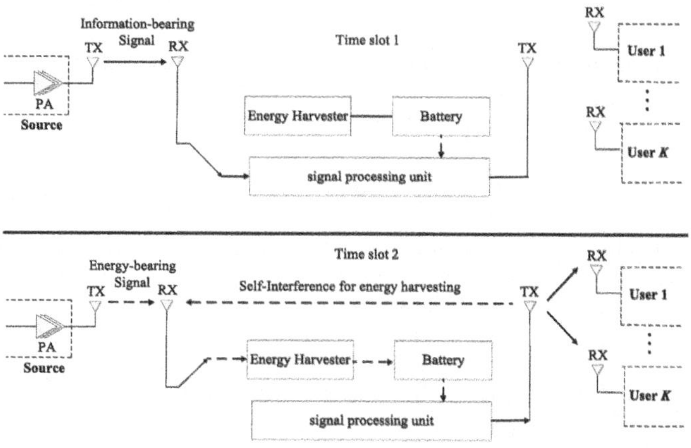

Fig. 3.1 Illustration of asymmetric WPT-aided FD DF relay multi-user system, where dashed and solid lines denote energy-bearing and information-bearing signals, respectively

element $\beta_{k,n} = \{1, 0\}$ denotes whether subcarrier n is assigned to user k (by 1) or not (by 0). In the first time slot, the received signal on subcarrier n for user k at the relay is calculated as

$$r_{k,n,1}[i] = \sqrt{p_{k,n} l_{sr}} h_{sr,n} x_{k,n}[i] + n_{r,n}[i] \quad (3.1)$$

where i represents the symbol index. $x_{k,n}[i]$ denotes the information symbol for user k on subcarrier n with normalized power $\mathbb{E}\{|x_{k,n}[i]|^2\} = 1$. $p_{s,k,n}$ denotes the transmission power allocated at the source for user k on subcarrier n. $h_{sr,n}$ denotes the channel response of the source-relay link on subcarrier n, and l_{sr} captures the large-scale fading of the source-relay link. $n_{r,n}[i]$ is the noise introduced on subcarrier n at the relay. Therefore, the overall SE of the source-relay link in the first time slot T_1 is calculated as

$$C_{SR} = \frac{T_1}{T} \sum_{k=1}^{K} \sum_{n=1}^{N} \beta_{k,n} \log_2(1 + p_{s,k,n} \gamma_{sr,k,n}), \quad (3.2)$$

where $\beta_{k,n}$ denotes subcarrier n is assigned for user k, $\forall k \in K, n \in N$. $\gamma_{sr,k,n} = \frac{|h_{sr,n}|^2 l_{sr}}{N_0 B}$ denotes the channel-to-noise ratio (CNR) of the source-relay link on subcarrier n for user k. N_0 is the single-sided power spectral density and B is the bandwidth of each subcarrier. In the second time slot of duration T_2, the received signal at the relay is forwarded to the user. The received signal by user k on subcarrier n is given by

$$y_{k,n}[i] = \sqrt{p_{r,k,n} l_{rd,k}} h_{rd,k,n} x_{k,n}[i] + z_{k,n}[i], \quad (3.3)$$

where $p_{r,k,n}$ denotes the transmission power for user k on subcarrier n at the relay. $h_{rd,k,n}$ is the channel response of the link between the relay and user k on subcarrier n for user k, and $l_{rd,k}$ captures the effects of large-scale fading of the relay-user k. $z_{k,n}[i]$ is the noise introduced at user k. Therefore, the overall SE in the second time slot is calculated as

$$C_{RD} = \frac{T_2}{T} \sum_{k=1}^{K} \sum_{n=1}^{N} \beta_{k,n} \log_2(1 + p_{r,k,n} \gamma_{rd,k,n}) \tag{3.4}$$

where $\gamma_{rd,k,n} = \frac{|h_{rd,n}|^2 l_{rd,k}}{N_0 B}$ denotes the CNR of the relay-user k on subcarrier n. Finally, the end-to-end SE in the whole time $T = T_1 + T_2$ is expressed as

$$C = \min\{C_{SR}, C_{RD}\} \tag{3.5}$$

Concurrently with data transmission by the relay in the second time slot T_2, wireless energy is sent from the source to the relay for the purpose of energy harvesting. Since data signal is transmitted by the relay at the same time, the relay also receives the self-interference from its own transmitter, which however can be collected as a green source for energy harvesting [76]. Hence, the received signal at the relay's receiver in the second time slot T_2 for user k on subcarrier n is calculated as

$$r_{k,n,2}[i] = \sqrt{p_{s,k,n} l_{sr}} h_{sr,n} x_{k,n}^{(e)}[i] + \sqrt{p_{r,k,n} l_{rr}} h_{rr,n} x_{k,n}[i - \tau] + n_{r,n}[i] \tag{3.6}$$

where $x_{k,n}^{(e)}[i]$ denotes the energy-bearing signal sent from the source to the relay in the second time slot T_2 with normalized energy $\mathbb{E}\{|x_{k,n}^{(e)}[i]|^2\} = 1$. τ denotes the processing delay at relay node. $h_{rr,n}$ denotes the self-interference channel from the relay's transmitter to its receiver on subcarrier n and l_{rr} denotes large-scale fading of the self-interference channel. The overall harvested energy at the relay during the second time slot T_2 is given as

$$E_r = \frac{T_2}{T} \omega \sum_{k=1}^{K} \sum_{n=1}^{N} \beta_{k,n} E\{|\sqrt{p_{s,k,n} l_{sr}} h_{sr,n} x_{k,n}^{(e)}[i] + \sqrt{p_{r,k,n} l_{rr}} h_{rr,n} x_{k,n}[i - \tau]|^2\}, \tag{3.7}$$

where ω denotes the energy harvesting efficiency [73, 79]. Since the channel condition and the transmitted signal $x_{s,k,n}^{(e)}[i]$ and $x_{s,k,n}[i - \tau]$ are known by the system, it is possible to write the state of $x_{s,k,n}^{(e)}[i]$ as

$$x_{s,k,n}[i] e^{j\angle h_{sr,n}} = x_{s,k,n}[i - \tau] e^{j\angle h_{rr,n}}, \tag{3.8}$$

where $\angle h_{sr,n}$ and $\angle h_{rr,n}$ denote the phase of the complex variables $h_{sr,n}$ and $h_{rr,n}$. In order to make the harvested energy maximized [76], we introduce the following Lemma 3.1.

Lemma 3.1 *In the asymmetric WPT FD relay OFDMA system, the maximal energy harvested at the relay is calculated as*

$$E_r = \frac{T_2}{T}\omega\sum_{k=1}^{K}\sum_{n=1}^{N}\beta_{k,n}(p_{s,k,n}|h_{sr,n}|^2 l_{sr} + p_{r,k,n}|h_{rr,n}|^2 l_{rr}). \tag{3.9}$$

3.1.2 Problem Formulation

Define $C(T_1, T_2, \boldsymbol{p_s}, \boldsymbol{p_r}, \boldsymbol{\beta})$ as the SE (in bps/Hz). Assume that the transmission power at the relay relies on WPT from the source node, while the circuit power consumed by the relay is negligible as compared to the transmission power. It which could be the case for small-sized and low-complexity devices (such as wireless sensors and tags) with simple electronics and low signal processing requirement [66, 76]. Hence, resource allocation optimization problem is formulated as

$$\begin{aligned}
P3.1: \quad & \underset{T_1, T_2, \boldsymbol{p_s}, \boldsymbol{p_r}, \boldsymbol{\beta}}{\text{argmax}} \; C(T_1, T_2, \boldsymbol{p_s}, \boldsymbol{p_r}, \boldsymbol{\beta}) = \\
& \min\{\frac{T_1}{T}\sum_{k=1}^{K}\sum_{n=1}^{N}\beta_{k,n}\log_2(1 + p_{s,k,n}\gamma_{sr,n}), \frac{T_2}{T}\sum_{k=1}^{K}\sum_{n=1}^{N}\beta_{k,n}\log_2(1 + p_{r,k,n}\gamma_{rd,k,n})\}, \\
& (C1): 0 \le p_{s,k,n}, \forall n \in N, k \in K. \\
& (C2): \sum_{k=1}^{K}\sum_{n=1}^{N}\beta_{k,n}p_{s,k,n} \le p_{max}, \\
& (C3): 0 \le p_{r,k,n}, \forall n \in N, k \in K. \\
& (C4): \sum_{k=1}^{K}\sum_{n=1}^{N}\beta_{k,n}p_{r,k,n} \le \omega\sum_{k=1}^{K}\sum_{n=1}^{N}\beta_{k,n}(p_{s,k,n}|h_{sr,n}|^2 l_{sr} + p_{r,k,n}|h_{rr,n}|^2 l_{rr}), \\
& (C5): T_1 + T_2 = T. \\
& (C6): \sum_{k=1}^{K}\beta_{k,n} = 1, \forall n \in N, \\
& (C7): \beta_{k,n} \in \{0, 1\}, \forall n \in N, k \in K,
\end{aligned} \tag{3.10}$$

where $(C1)$ and $(C2)$ denote that the transmission power at the source should be non-negative and lower than a budget p_{max}. $(C3)$ and $(C4)$ are the constraints imposed on transmission power at the relay, which should be non-negative and lower than the total harvested power. $(C5)$ indicates that the summation of two time slots T_1 and T_2 should be equivalent to T. $(C6)$ and $(C7)$ guarantee that each subcarrier is only used by one user. The allocation of time slots T_1 and T_2, subcarrier $\boldsymbol{\beta}$, transmission power $\boldsymbol{p_s}$ and $\boldsymbol{p_r}$ need to be optimized jointly, subject to multiple constraints $(C1) - (C7)$.

3.2 Transformation of the Optimization Problem

To solve the non-convex problem P3.1, the min function in the objective function should be handled first. It can be deduced that P3.1 is maximized only when

$$\frac{T_1}{T}\sum_{k=1}^{K}\sum_{n=1}^{N}\beta_{k,n}\log_2(1+p_{s,k,n}\gamma_{sr,k,n}) = \frac{T_2}{T}\sum_{k=1}^{K}\sum_{n=1}^{N}\beta_{k,n}\log_2(1+p_{r,k,n}\gamma_{rd,k,n}), \tag{3.11}$$

which is easy to prove by a counter example. Therefore, the original optimization problem

$$\min\{\frac{T_1}{T}\sum_{k=1}^{K}\sum_{n=1}^{N}\beta_{k,n}\log_2(1+p_{s,k,n}\gamma_{sr,n}), \frac{T_2}{T}\sum_{k=1}^{K}\sum_{n=1}^{N}\beta_{k,n}\log_2(1+p_{r,k,n}\gamma_{rd,k,n})\} \tag{3.12}$$

is equivalent to maximizing the following objective function

$$\frac{T_1}{T}\sum_{k=1}^{K}\sum_{n=1}^{N}\beta_{k,n}\log_2(1+p_{s,k,n}\gamma_{sr,k,n}) + \frac{T_2}{T}\sum_{k=1}^{K}\sum_{n=1}^{N}\beta_{k,n}\log_2(1+p_{r,k,n}\gamma_{rd,k,n}), \tag{3.13}$$

with an additional constraint

$$\frac{T_1}{T}\sum_{k=1}^{K}\sum_{n=1}^{N}\beta_{k,n}\log_2(1+p_{s,k,n}\gamma_{sr,k,n}) = \frac{T_2}{T}\sum_{k=1}^{K}\sum_{n=1}^{N}\beta_{k,n}\log_2(1+p_{r,k,n}\gamma_{rd,k,n}). \tag{3.14}$$

Then the combinational constraint $(C7)$ caused by the subcarrier assignment is relaxed. The binary variable $\beta_{k,n}$, for $\forall k \in K, n \in N$, is relaxed to a real number within the interval [0,1], indicating the time sharing factor of subcarrier n among K

users. Given a large number of subcarriers, the loss of optimality is negligible and the solution after relaxation is near-optimal [28]. Hence, the transformed problem P3.2 is given by (3.15).

$$
\begin{aligned}
P3.2: \quad & \underset{T_1, T_2, \boldsymbol{p_s}, \boldsymbol{p_r}, \boldsymbol{\beta}}{\operatorname{argmax}} \quad C(T_1, T_2, \boldsymbol{p_s}, \boldsymbol{p_r}, \boldsymbol{\beta}) = \\
& \frac{T_1}{T}\sum_{k=1}^{K}\sum_{n=1}^{N}\beta_{k,n}\log_2(1+\frac{p_{s,k,n}\gamma_{sr,k,n}}{\beta_{k,n}}) + \frac{T_2}{T}\sum_{k=1}^{K}\sum_{n=1}^{N}\beta_{k,n}\log_2(1+\frac{p_{r,k,n}\gamma_{rd,k,n}}{\beta_{k,n}}), \\
& s.t.\ (C1), (C3), (C5), (C6), \\
& (\widetilde{C2}): \sum_{k=1}^{K}\sum_{n=1}^{N} p_{s,k,n} \le p_{max}, \\
& (\widetilde{C4}): \sum_{k=1}^{K}\sum_{n=1}^{N} p_{r,k,n} \le \omega\sum_{k=1}^{K}\sum_{n=1}^{N}(p_{s,k,n}|h_{sr,n}|^2 l_{sr} + p_{r,k,n}|h_{rr,n}|^2 l_{rr}), \\
& (\widetilde{C7}): \beta_{k,n} \in [0,1], \forall n \in N, k \in K, \\
& (C8): \frac{T_1}{T}\sum_{k=1}^{K}\sum_{n=1}^{N}\beta_{k,n}\log_2(1+\frac{p_{s,k,n}\gamma_{sr,k,n}}{\beta_{k,n}}) = \\
& \qquad \frac{T_2}{T}\sum_{k=1}^{K}\sum_{n=1}^{N}\beta_{k,n}\log_2(1+\frac{p_{r,k,n}\gamma_{rd,k,n}}{\beta_{k,n}})
\end{aligned} \tag{3.15}
$$

The transformation from $P3.1$ to $P3.2$ is straightforward, and it should be noticed that P3.2 is subject to a new constraint (C8). Now Theorem 3.1 is given to solve the optimization problem P3.2.

Theorem 3.1 *The equivalent end-to-end SE* $\frac{T_1}{T}\sum_{k=1}^{K}\sum_{n=1}^{N}\beta_{k,n}\log_2(1+\frac{p_{s,k,n}\gamma_{sr,k,n}}{\beta_{k,n}})+\frac{T_2}{T}\sum_{k=1}^{K}\sum_{n=1}^{N}\beta_{k,n}\log_2(1+\frac{p_{r,k,n}\gamma_{rd,k,n}}{\beta_{k,n}})$ *is jointly-concave with respect to all variables* T_1, T_2, $p_{s,k,n}$, $p_{r,k,n}$, $\beta_{k,n}$, *for* $\forall n \in N, k \in K$ *in the feasible domain.*

The proof of theorem 3.1 is provided in Appendix F.

3.3 Solution to the Asymmetric Resource Allocation for WPT-aided FD Relay Multiuser Systems

Because P3.2 is a standard concave optimization problem, it can be solved by using the KKT conditions. Its Lagrangian function is given by

$$
\begin{aligned}
L = &\frac{T_1}{T}\sum_{k=1}^{K}\sum_{n=1}^{N}\beta_{k,n}\log_2(1+\frac{p_{s,k,n}\gamma_{sr,k,n}}{\beta_{k,n}})\\
&+\frac{T_2}{T}\sum_{k=1}^{K}\sum_{n=1}^{N}\beta_{k,n}\log_2(1+\frac{p_{r,k,n}\gamma_{rd,k,n}}{\beta_{k,n}})\\
&-\mu(\sum_{k=1}^{K}\sum_{n=1}^{N}p_{s,k,n}-p_{max})\\
&-\phi(\sum_{k=1}^{K}\sum_{n=1}^{N}p_{r,k,n}(1-\omega|h_{rr,n}|^2 l_{rr})-\rho(T_1+T_2-T)\\
&-\lambda(\frac{T_1}{T}\sum_{k=1}^{K}\sum_{n=1}^{N}\beta_{k,n}\log_2(1+\frac{p_{s,k,n}\gamma_{sr,k,n}}{\beta_{k,n}})\\
&-\frac{T_2}{T}\sum_{k=1}^{K}\sum_{n=1}^{N}\beta_{k,n}\log_2(1+\frac{p_{r,k,n}\gamma_{rd,k,n}}{\beta_{k,n}}))\\
&-\omega(\sum_{k=1}^{K}\sum_{n=1}^{N}p_{s,k,n}|h_{sr,n}|^2 l_{sr}))-\sum_{n=1}^{N}\psi_n(\sum_{k=1}^{K}\beta_{k,n}-1),
\end{aligned}
\tag{3.16}
$$

where μ, ϕ, ρ, λ and ψ_n (for $\forall n \in N$) are Lagrange multipliers. It can be derived from the KKT conditions that $\mu \geq 0, \phi \geq 0, \rho \geq 0, \lambda \in (-1, 1)$ and $\psi_n \geq 0$ (for $\forall n \in N$). Let $\mu^*, \phi^*, \rho^*, \lambda^*$ and ψ_n^* (for $\forall n \in N$) denote the optimal Lagrange multipliers, it has

$$
\frac{\partial L}{\partial T_1}=\frac{(1-\lambda^*)}{T}\sum_{k=1}^{K}\sum_{n=1}^{N}\beta_{k,n}^*\log_2(\frac{1+p_{s,k,n}^*\gamma_{sr,k,n}}{\beta_{k,n}^*})-\rho^*\begin{cases}<0, T_1^*=0\\=0, T_1^*>0\end{cases}
\tag{3.17}
$$

$$
\frac{\partial L}{\partial T_2}=\frac{(1+\lambda^*)}{T}\sum_{k=1}^{K}\sum_{n=1}^{N}\beta_{k,n}^*\log_2(1+\frac{p_{r,k,n}^*\gamma_{rd,k,n}}{\beta_{k,n}^*})-\rho^*\begin{cases}<0, T_2^*=0\\=0, T_2^*>0\end{cases}
\tag{3.18}
$$

From (3.17) and (3.18) and constraint (C8), an equality can be obtained as

$$
\frac{T_1^*}{1-\lambda^*}=\frac{T_2^*}{1+\lambda^*},
\tag{3.19}
$$

According to (3.19) and constraint (C5), it can be derived that the optimal time slot T_1 is given by

$$T_1^* = \frac{1-\lambda^*}{2}T. \tag{3.20}$$

and the optimal time slot T_2^* is

$$T_2^* = \frac{1+\lambda^*}{2}T. \tag{3.21}$$

Also, taking the derivative of L with respect to $p_{s,k,n}$ and $p_{r,k,n}$ yields

$$\frac{\partial L}{\partial p_{s,k,n}} = \frac{(1-\lambda^*)T_1^*\gamma_{sr,k,n}\beta_{k,n}^*}{T(\beta_{k,n}^* + p_{s,k,n}^*\gamma_{sr,k,n})} - \mu^* + \phi^*\omega|h_{sr,k,n}|^2 l_{sr} \begin{cases} < 0, \ p_s^* = 0 \\ = 0, \ p_s^* > 0 \end{cases} \tag{3.22}$$

$$\frac{\partial L}{\partial p_{r,k,n}} = \frac{(1+\lambda^*)T_2^*\gamma_{rd,k,n}\beta_{k,n}^*}{T(\beta_{k,n}^* + p_{r,k,n}^*\gamma_{rd,k,n})} - \phi^*(1-\omega|h_{rr,n}|^2 l_{rr}) = \begin{cases} < 0, \ p_r^* = 0 \\ = 0, \ p_r^* > 0 \end{cases} \tag{3.23}$$

Substitute (3.20) and (3.21) into (3.22) and (3.23), and then the optimal power allocation at the source is calculated as

$$p_{s,k,n}^* = [\frac{(1-\lambda^*)^2\beta_{k,n}^*}{2(\mu^* - \phi^*\omega|h_{sr,n}|^2 l_{sr})} - \frac{\beta_{k,n}^*}{\gamma_{sr,k,n}}]^+, \tag{3.24}$$

where $[x]^+ = max\{0, x\}$. Also, the optimal power allocation at the relay is calculated as

$$p_{r,k,n}^* = [\frac{(1+\lambda^*)^2\beta_{k,n}^*}{2\phi^*(1-\omega|h_{rr,n}|^2 l_{rr})} - \frac{\beta_{k,n}^*}{\gamma_{rd,k,n}}]^+, \tag{3.25}$$

By taking derivative L with respect to $\beta_{k,n}$, we have

$$\begin{aligned}\frac{\partial L}{\partial \beta_{k,n}} &= \frac{(1-\lambda^*)T_1}{T}\log_2(1+\frac{p_{s,k,n}^*\gamma_{sr,k,n}}{\beta_{k,n}^*}) - \frac{(1-\lambda^*)T_1^* p_{s,k,n}^*\gamma_{sr,k,n}}{T(\beta_{k,n}^* + p_{s,k,n}^*\gamma_{sr,k,n})} \\ &+ \frac{(1+\lambda^*)T_2^*}{T}\log_2(1+\frac{p_{r,k,n}^*\gamma_{rd,k,n}}{\beta_{k,n}^*}) - \frac{(1+\lambda^*)T_2^* p_{r,k,n}^*\gamma_{rd,k,n}}{T(\beta_{k,n}^* + p_{r,k,n}^*\gamma_{rd,k,n})} - \psi_n^* \\ &= \begin{cases} < 0, \ \beta_{k,n}^* = 0 \\ \geq 0, \ \beta_{k,n}^* = 1. \end{cases}\end{aligned} \tag{3.26}$$

Since $\beta_{k,n}$ is either 0 or 1, by substituting (3.20), (3.21), (3.24) and (3.25) into (3.26), it is obtained that the subcarrier n will be assigned to the user k satisfying

$$\beta_{k,n}^* = \begin{cases} 1, k = k^* \\ 0, k \neq k^*, \end{cases} \tag{3.27}$$

where we have

$$\begin{aligned} k^* = \underset{k}{\operatorname{argmax}} & \frac{(1-\lambda^*)^2}{2}[\log_2(\frac{(1-\lambda^*)^2}{2}\frac{\gamma_{sr,k,n}}{(\mu^* - \phi^*\omega|h_{sr,n}|^2 l_{sr})})]^+ \\ & + \frac{(1+\lambda^*)^2}{2}[\log_2(\frac{(1+\lambda^*)^2}{2}\frac{\gamma_{rd,k,n}}{\phi^*(1-\omega|h_{rr,n}|^2 l_{rr})})]^+ \\ & - [\frac{(1-\lambda^*)^2\gamma_{sr,k,n} - 2(\mu^* - \phi^*\omega|h_{sr,n}|^2 l_{sr})}{2\gamma_{sr,k,n}}]^+ \\ & - [\frac{(1+\lambda^*)^2\gamma_{rd,k,n} - 2\phi^*(1-\omega|h_{rr,n}|^2 l_{rr})}{2\gamma_{rd,k,n}}]^+ \end{aligned} \tag{3.28}$$

Hence, the optimal value of $p_{s,k,n}$ and $p_{r,k,n}$ is finally given by

$$p_{s,k,n}^* = [\frac{(1-\lambda^*)^2}{2(\mu^* - \phi^*\omega|h_{sr,n}|^2 l_{sr})} - \frac{1}{\gamma_{sr,k,n}}]^+. \tag{3.29}$$

$$p_{r,k,n}^* = [\frac{(1+\lambda^*)^2}{2\phi^*(1-\omega|h_{rr,n}|^2 l_{rr})} - \frac{1}{\gamma_{rd,k,n}}]^+. \tag{3.30}$$

After obtaining the optimal results T_1^*, T_2^*, $p_{s,k,n}^*$, $p_{r,k,n}^*$ and $\beta_{k,n}^*$ (for $\forall k \in K$ and $\forall n \in N$), the gradient method can be used to update the multipliers μ, ϕ, λ and ρ, while multiplier ψ_n (for $\forall n \in N$) has been omitted in deriving (3.27). Based on the gradient method, the updating rules can be given as

$$\begin{aligned}
&\mu(j+1)=[\mu(j)+\epsilon_1(\sum_{k=1}^{K}\sum_{n=1}^{N}p_{s,k,n}-p_{max})]^+;\\
&\phi(j+1)=[\phi(j)+\epsilon_2(\sum_{k=1}^{K}\sum_{n=1}^{N}p_{r,k,n}(1-\omega|h_{rr,n}|^2l_{rr})\\
&\quad-\omega\sum_{k=1}^{K}\sum_{n=1}^{N}p_{s,k,n}|h_{sr,n}|^2l_{sr})]^+;\\
&\lambda(j+1)=[\lambda(j)+\epsilon_3(\sum_{k=1}^{K}\sum_{n=1}^{N}\beta_{k,n}\log_2(1+\frac{p_{s,k,n}\gamma_{sr,k,n}}{\beta_{k,n}})-\\
&\quad\sum_{k=1}^{K}\sum_{n=1}^{N}\beta_{k,n}\log_2(1+\frac{p_{r,k,n}\gamma_{rd,k,n}}{\beta_{k,n}}))]^+;\\
&\rho(j+1)=[\rho(j)+\epsilon_4(T_1+T_2-T)]^+;
\end{aligned}\tag{3.31}$$

where ϵ_1, ϵ_2 ϵ_3, and ϵ_4 are the corresponding step sizes [28]. Now the optimisation problem P3.2 is readily solved. Based on our theoretic analysis, a so-called asymmetric WPT-aided FD relay (Asym-WPT-FR) algorithm is proposed in Algorithm 3.

Algorithm 3 Asymmetric WPT-aided FD relay (Asym-WPT-FR) algorithm

Input: Initial value of multipliers $\mu(0)$, $\phi(0)$, $\lambda(0)$ and $\rho(0)$, channel condition $h_{sr,n}$, $h_{rd,k,n}$, $h_{rr,n}$, l_{sr}, $l_{rd,k}$, and l_{rr}, power related parameters p_{max} and ω.
Output: Optimal Lagrange multipliers λ^*, ϕ^*, μ^* and ρ^*, optimal results T_1^*, T_2^*, $p_{s,k,n}^*$, $p_{r,k,n}^*$ and $\beta_{k,n}^*$, for $\forall n \in N, k \in K$.
repeat
  Allocate subcarrier n to user k satisfying (21).
  Allocate T_1, T_2, p_s, p_r as
  $T_1=\frac{1-\lambda}{2}T$;
  $T_2=\frac{1+\lambda}{2}T$;
  $p_{s,k,n}=[\frac{(1-\lambda)^2}{2(\mu-\phi\omega|h_{sr,n}|^2l_{sr})}-\frac{1}{\gamma_{sr,k,n}}]^+$;
  $p_{r,k,n}=[\frac{(1+\lambda)^2}{2\phi(1-\omega|h_{rr,n}|^2l_{rr})}-\frac{1}{\gamma_{rd,k,n}}]^+$;
  Update multipliers
  $\mu(j+1)=[\mu(j)+\epsilon_1(\sum_{k=1}^{K}\sum_{n=1}^{N}p_{s,k,n}-p_{max})]^+$;
  $\phi(j+1)=[\phi(j)+\epsilon_2(\sum_{k=1}^{K}\sum_{n=1}^{N}p_{r,k,n}(1-\omega|h_{rr,n}|^2l_{rr})-\omega\sum_{k=1}^{K}\sum_{n=1}^{N}p_{s,k,n}|h_{sr,n}|^2l_{sr})]^+$;
  $\lambda(j+1)=[\lambda(j)+\epsilon_3(\sum_{k=1}^{K}\sum_{n=1}^{N}\beta_{k,n}\log_2(1+\frac{p_{s,k,n}\gamma_{sr,k,n}}{\beta_{k,n}})-\sum_{k=1}^{K}\sum_{n=1}^{N}\beta_{k,n}\log_2(1+\frac{p_{r,k,n}\gamma_{rd,k,n}}{\beta_{k,n}}))]^+$;
  $\rho(j+1)=[\rho(j)+\epsilon_4(T_1+T_2-T)]^+$;
until Convergence

3.4 Complexity Analysis and Properties of WPT FD Relay Based Multiuser Systems

The computational complexity of the algorithm is expressed as $\mathcal{O}(NK \times l_{ite})$, where l_{ite} denotes the number of iterations for updating multipliers until convergence [57]. In the followings, we provide some properties regarding optimal resource allocation in asymmetric WPT FD relay systems.

3.4.1 *Length of Time Slots in WPT FD DF Relay Systems*

In asymmetric WPT FD relay systems, the length of time slot T_2 is generally longer than that of time slot T_1. It is because the relay's power is harvested from the source, which is much lower than the transmission power at the source due to the signal attenuation from the source to the relay and low harvesting efficiency. As a result, the SINR of the source-relay link overwhelms the SINR of the relay-users link, which confines the value of λ into (0,1), as can be seen from (3.31). Hence, the value $T_2 = \frac{1+\lambda}{2}T$ is higher than that of $T_1 = \frac{1-\lambda}{2}T$. In some extreme cases where the distance of the relay-destination is much shorter than that of the source-relay, the SINR of the relay-user link may be higher than that of the source-relay link. The multiplier λ in this case may fall into the domain $(-1,0)$ and thus T_1 may be higher than T_2 to balance the SE performance of the two links. In a nutshell, based on the SINR values of the two links, the asymmetric system can dynamically adjust the lengths of T_1 and T_2 to optimize the overall system SE.

3.4.2 *Power Allocation Design in WPT FD DF Relay Systems*

As can be seen from (3.30), the transmission power at the relay node is jointly determined by the channel frequency responses $|h_{rr,k,n}|$ and $|h_{rd,k,n}|$. With the same channel response on $|h_{rd,k,n}|$, it shows that the proposed Asym-WPT-FR system prefers to assign more power on the subcarrier with a higher value of $|h_{rr,k,n}|$. On the contrary, more transmission power is allocated on the subcarrier with a lower value of $|h_{rr,k,n}|$ in conventional FD grid-powered relay systems [20]. It is because in the considered WPT relay systems, self-interference is used as a green source for energy harvesting and thus more energy can be harvested with a higher value of $|h_{rr,k,n}|$. By contrast, as self-interference is treated as noise in [20, 80], a low channel frequency response of the self-interference channel is desirable.

3.5 Performance Comparison

The performance of the proposed Asym-WPT-FR system is demonstrated in this section, compared with the performance of the following systems:

(1) Sym-WPT-FR system [76], where the two time slots are of equal length. (2) TS-WPT-FR system [58], where the relay only harvests energy from the source in the first time slot. In the second time slot, the source transmits signal to the destinations with the help of a FD relay. The self-interference is mitigated as noise at the relay node, and amount of the self-interference cancelled in [58] is set to 80 dB. The values of the time slot durations, T_1 and T_2, are calculated according to [58, Eq.(16)].

The number of subcarrier is set to $N = 128$ with per-subcarrier bandwidth $B =$ 156 kHz. The number of users K is set to 3 except for Fig. 3.4. The power spectral density of AWGN is $N_0 = -174$ dBm/Hz. The PL model for a practical scenario at 900 MHz [66] is used as PL $= 31.7 + 10\theta \log_{10}(d/d_0)$, where $\theta = 2.5$ denotes the PL exponent, d denotes the distance between two communication nodes, and $d_0 = 1$ m denotes the reference distance. Without loss of generality, it is set that $d_{SD} = d_{SR} + d_{RD} = 100$ m, where d_{SD}, d_{SR} and d_{RD} are the distances of the source-users, the source-relay and the relay-users links. A six-path Rayleigh fading channel with an exponential delay profile and a RMS delay spread of 0.022 us are used. The energy harvesting efficiency of the WPT relay node ω is set to 0.8.

3.5.1 SE Performance

Figure 3.2 shows the SE performance with different levels of transmission power at the source. It is demonstrated that the Asym-WPT-FR achieves around twice higher SE than the Sym-WPT-FR system. It is because the SE of the two links

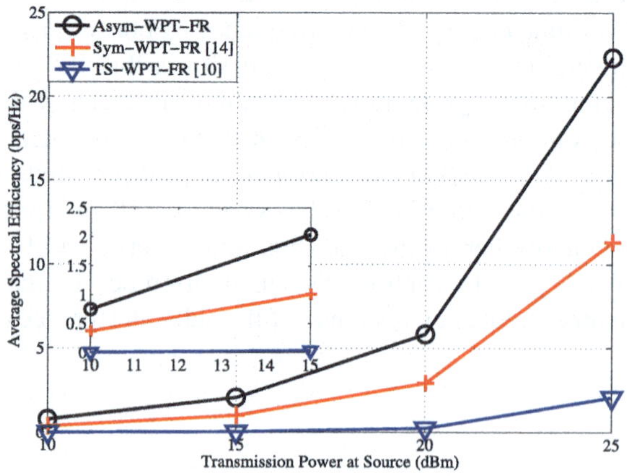

Fig. 3.2 Impact of the transmission power at the source on SE performance with $d_{SR} = d_{RD} = 50$ m

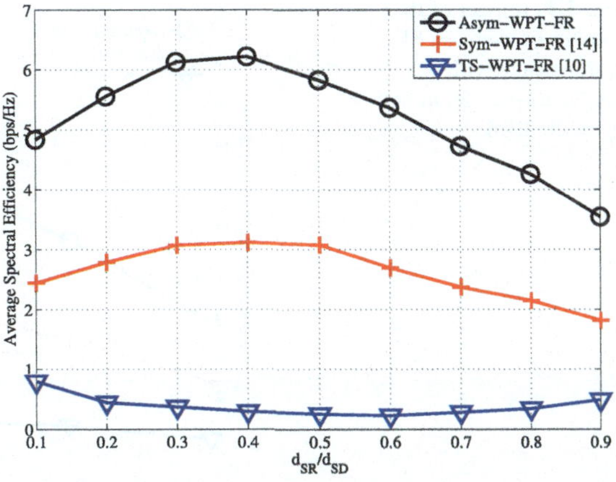

Fig. 3.3 Impact of the normalized source-relay distance on SE performance with $d_{SD} = 100$ m and $p_s = 20$ dBm

can be balanced by the Asym-WPT-FR system, benefiting from the higher degree of freedom of asymmetric time slots. By contrast, the Sym-WPT-FR system allocates uniform time slots for its source and relay, whereas the effective SINR of the relay-user link is much lower than that of the source-relay link. As a result, the end-to-end SE is bounded by the relay-user link, and the SE of the source-relay link is wasted partially. The TS-WPT-FR shows the poorest SE performance among the three systems. It is because its relay's transmission power is only harvested from the source in the first time slot, and self-interference is not effectively utilised, resulting in a low available transmission power at the relay node. Therefore, the SE of the TS-WPT-FR system is limited by its low transmission power.

Figure 3.3 demonstrates the SE performance with different relay's positions. The proposed Asym-WPT-FR system shows the highest SE among the three systems, showing the robustness on the relay's position. With different relay's position, the SINRs of the source-relay and the relay-user vary, and the Asym-WPT-FR system can adjust the lengths of two time slots to equate the SE of the two hops. Besides, higher SE can be achieved when the relay moves to the middle point between the source and the users, which is because the relay can harvest reasonable energy from source and is not so far from the users around the middle point. While the SE of the TS-WPT-FR system is maximised when the relay is either close to the source or close to the users. It is because in the TS-WPT-FR system, T_2 (used for information transmission only) approaches 40% in the aforementioned two cases and a higher SE is obtained.

Figure 3.4 shows the SE performance with different numbers of users. It is demonstrated that the SE of the Asym-WPT-FR system increases faster than others. It is because the user diversity of the relay-multiuser link increases with more users, which helps improve the system SE. While the TS-WPT-FR and the Sym-WPT-FR have little increase in SE with the increase of the number of users.

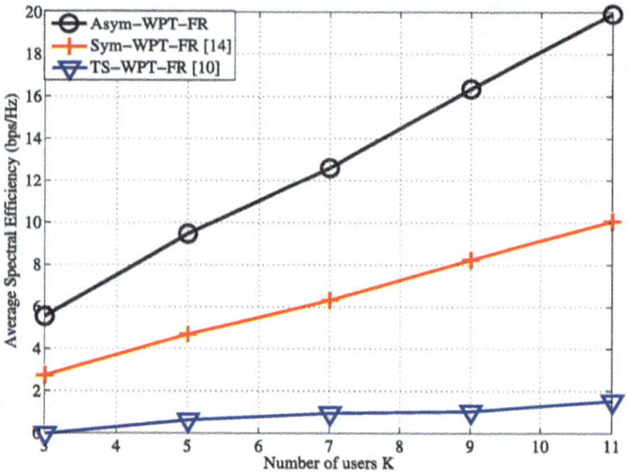

Fig. 3.4 Impact of the number of users on SE performance with $d_{SD} = d_{RD} = 50$ m and $p_s = 20$ dBm

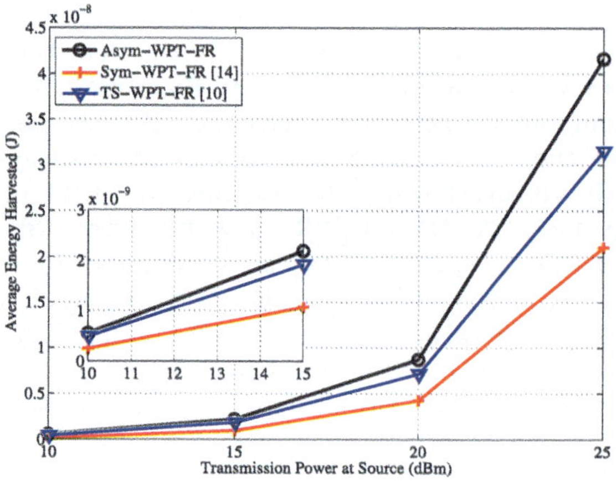

Fig. 3.5 Impact of the transmission power on the energy harvested with $d_{SR} = d_{RD} = 50$ m

3.5.2 *Harvested Energy*

Figure 3.5 demonstrates the harvested energy with different levels of transmission power. It is observed that the proposed Asym-WPT-FR system harvests the most energy, achieving at least 30% enhancement over the Sym-WPT-FR system. Since the asymmetric system can provide more time for the second time slot T_2 to compensate the weak SINR at user side, allowing the relay to have more time to harvest energy. Besides, in the Asym-WPT-FR system, energy comes from both source and relay. As a comparison, the TS-WPT-FR system only harvests energy from the source, therefore it shows the lowest harvested energy, only a half of the energy harvested by our Asym-WPT-FR system.

Figure 3.6 demonstrates the amount of the harvested energy with different relay positions. The Asym-WPT-FR system can harvest more energy than others. Also, it is obvious that the harvested energy decreases given a longer source-relay distance due to the signal transmission attenuation.

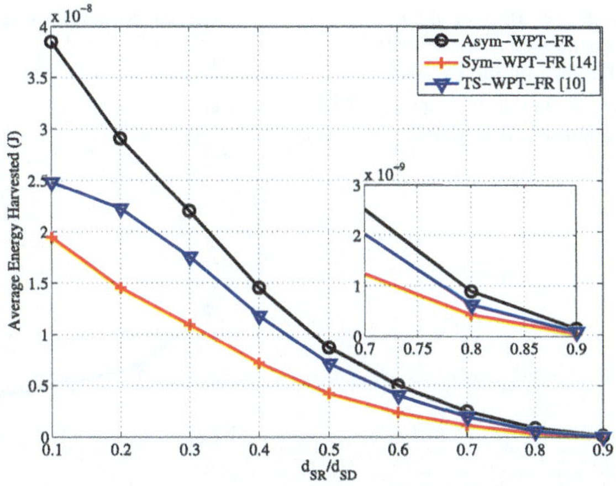

Fig. 3.6 Impact of the normalised source-relay distance on the energy harvested with $d_{SD} = 100$ m and $p_s = 20$ dBm

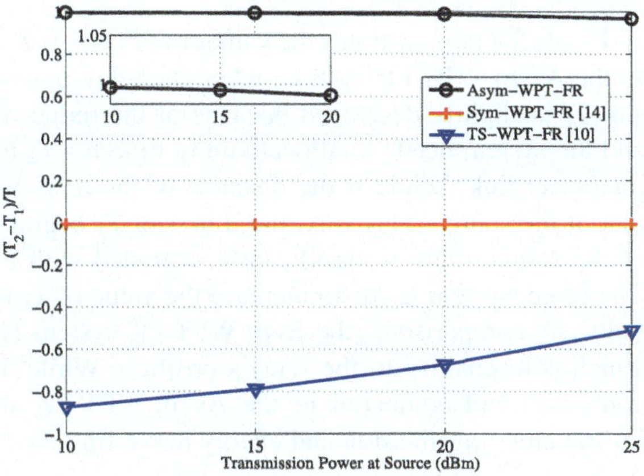

Fig. 3.7 Impact of the transmission power on the value of $(T_2 - T_1)/T$ with $d_{SR} = d_{RD} = 50$ m

3.5.3 Time Duration

Figure 3.7 shows the value of $(T_2 - T_1)/T$ under different levels of transmission power at the source. In the Asym-WPT-FR system, the length of the first time slot T_1 can be slightly increased with a higher transmission power at source. It is because with a higher transmission power at the source, more energy at the relay can be harvested and the SINR of the relay-user link is accordingly enhanced. System needs to reduce the length of the second time slot T_2 and replenish it to the first time slot T_1. On the contrary, T_1 decreases with a higher transmission power at the source in the TS-WPT-FR system. It is because time slot T_1 is only used for energy harvesting. Wireless energy may be collected quickly with a higher transmission power, and the system can allocate more time to the second time slot T_2 for data transmission. While Sym-WPT-FR system allocates symmetric time slots, and thus its SE of the source-relay link is wasted partially.

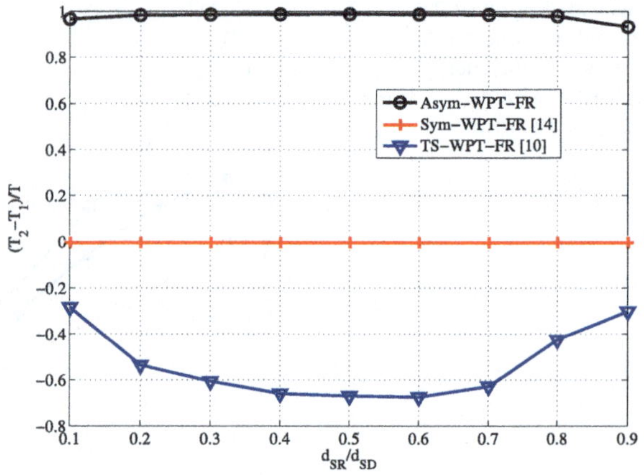

Fig. 3.8 Impact of the normalised source-relay distance on the value of $(T_2 - T_1)/T$ with $d_{SD} = 100$ m and $p_s = 20$ dBm

Figure 3.8 demonstrates the values of $(T_2 - T_1)/T$ with different relay's positions. In the Asym-WPT-FR system, when the relay moves to the middle point, the time slot T_1 is slightly decreased because of the reduced harvested power at the relay, and the system needs to allocate more time for T_2 to remedy the low SINR of the relay-user link. While if the distance of the relay-users link is much shorter than that of the source-relay link, the time slot T_1 begins to improve because the SINR of the source-relay is significantly degraded with a longer source-relay distance. Therefore, system needs to increase the value of T_1 to re-balance the SE of the two links. As comparisons, the Sym-WPT-FR system has uniform time slot duration, which is insensitive to the relay's position. While the TS-WPT-FR system shows converse trend compared to the Asym-WPT-FR algorithm, due to the different information transmission and energy harvesting mechanism.

Chapter 4
Energy-Efficient Full-Duplex Cooperative Non-orthogonal Multiple Access

Non-orthogonal multiple access (NOMA) techniques have received considerable attention due to the increased multiuser access capability. The key advantage of NOMA is that of allowing multiple users to share the same frequency-, time- or code-resource [81]. For users having better channel conditions, which are refereed to as strong users (SU)s, they can perform successive interference cancellation to decode the signal iteratively. For the users having poor channel conditions, which are refereed to as weak users (WU)s, they need to decode their own signal directly. Classic NOMA designs can be classified as non-cooperative systems, as there is no assistance between the SUs and WUs. In [82], the authors considered a two-user NOMA system at downlink, and studied the power allocation design to maximize the total SE. Based on the KKT conditions, a closed form of optimal result is calculated for the non-cooperative NOMA system. In [83], the authors proposed a hybrid NOMA system, where NOMA and OMA can be dynamically switched to achieve a higher SE. However, the power allocation factor for the SU and WU in [83] is pre-determined and is dependent from the instantaneous channel conditions, leading to SE loss. In [84], outage probabilities were researched in non-cooperative NOMA system with imperfect channel estimation. At user end, due to the channel estimation error, successive interference cancellation operation is not perfect, which further leads to error propagation problem. In [85], the authors studied the power and channel allocation designs for NOMA system. Based on the Lagrange dual programming, a sub-optimal algorithm is proposed to maximize the weighted overall SER. In [86] and [87], fairness issue in NOMA systems was researched, and a max-min fairness optimization was proposed to allocate power for multiple users. In [88, 89], MIMO NOMA system was investigated in terms of outage probability and SE. It was demonstrated that by integrating MIMO and NOMA, inter-cluster interference can be mitigated. In [90], NOMA with fixed power allocation was researched, where capacity gain of NOMA outperforming OMA was derived. It was found that, to highlight the advantage of NOMA, it is better to schedule two users whose channel conditions are distinctive into one group. In [91], NOMA was treated

Z. Wei et al., *Energy Efficient Full Duplex Wireless Communication Systems*, Wireless Networks, https://doi.org/10.1007/978-981-97-8915-3_4

as a special case of cognitive radio system. A SU with strong channel condition is viewed as a secondary user, while a WU is viewed as a primary user. Therefore, the main problem of [91] is to allocate power to secondary user on the condition that performance of the primary user will not be degraded too much.

For SUs, since the WU's signal imposes limited interference on the stronger signal, the WU's signal can be detected first and subtracted from the composite signals. Then, the resultant symbols are re-modulated and then be forwarded to the WUs. Such a system is called as cooperative NOMA. The throughput and outage performance of cooperative NOMA was investigated in [92, 93], where the BS broadcasts superimposed signals to both the SU and the WU in the first time slot, while the SU forwards the WU's signal to the WU in the second time slot. However, such a cooperative NOMA assume that the SU operates in HD mode and hence an additional time slot is required for relaying WU's data, which leads to SE loss. As an improvement, FD can be introduced into cooperative NOMA to avoid performance loss. By applying SIC at the SU, the throughput of FD cooperative NOMA potentially doubles that of HD cooperative NOMA [94–98].

Nevertheless, there are some fundamental challenges in cooperative NOMA systems: (a) The signal-to-interference plus noise-ratio (SINR) experienced at the WU can indeed be enhanced. However, the WU's SINR observed at the SU is reduced by the residual SI caused by the FD operation. As a result, the SINR difference between the SU and the WU is reduced and may even become negative. Then the original SU becomes the new WU and hence fails to perfectly decode the original WU's signal. This undesired phenomenon can be regarded the as "SINR gap reversal" problem, which becomes more severe when the relaying power at the SU keeps on increasing, as shown in Fig. 4.1. (b) In cooperative NOMA systems, additional transmission power and transmit circuit power are required for relaying the WU's signal. Hence, EE issue becomes a critical challenge. In contrast to conventional throughput maximization design, it is more beneficial to strike a trade-off between the throughput and EE. EE maximization [99] and power

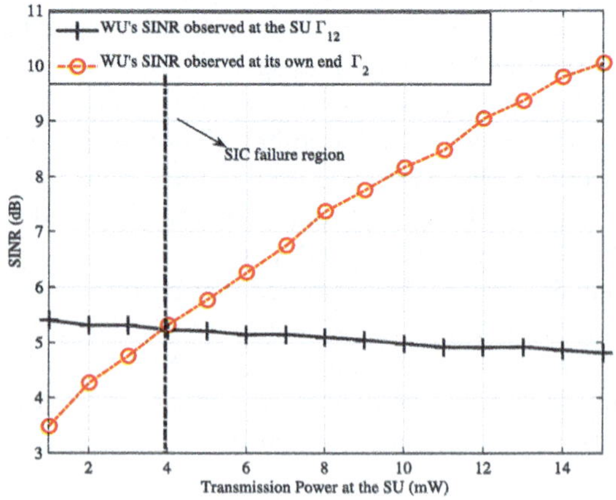

Fig. 4.1 The illustration of the "SINR gap reversal" in cooperative NOMA. To be specific, when the relaying power at the SU is high enough, the SU may fail to decode the WU's signal due to the "SINR gap reversal"

minimization [100] have been analyzed in non-cooperative NOMA, where the overwhelming majority of the transmission power is assigned to the distant WU for maintaining its SINR target. As a result, low throughput and EE are achieved due to the poor channel conditions between the BS and the WU. (c) The FD cooperative NOMA systems of [95, 96] assumed non-adaptive power allocation schemes, which are unaware of the users' channel conditions and suffer from significant throughput loss. More explicitly, given a large distance between the SU and WU, assigning high transmission power to the SU may not substantially improve the WU's throughput, while significantly degrading the EE of the SU.

4.1 System Model and Problem Formulation

We consider a classic NOMA system supporting two users by a BS. The users have different channel conditions on the same frequency, which is a common scenario in NOMA systems. The BS, SU and WU are equipped with a single antenna for the sake of having a low hardware complexity. In particular, the SU acts as a DF relay for forwarding the WU's signal. FD operation is enabled at the SU, which can transmit and receive signals at the same time [95]. The CSI is obtained by channel estimation [99]. To be specific, in the training phase, the BS estimates the downlink CSI by inferring it from the channel quality indicator feedback received from the SU and WU. Hence, the BS can update the CSI of h_1 and h_2. Furthermore, since the SU is aware of its own channel quality indicator and of the WU, the CSI of h_r and h_{12} can also be reported to the BS. Since a three-node single-antenna scenario is assumed, the signaling overhead (power, bandwidth, and throughput) maintains low [101, 102].

We define the transmission power allocated to the SU and to the WU at the BS by p_1 and p_2, where the the transmission power budget at the BS is denoted by $p_{BS,max}$. Let us furthermore define the transmission power consumed at the SU by p_3, which is upper bounded by the maximum available transmission power constraint $p_{3,max}$. Furthermore, to ensure that the SU can successfully decode the WU's signal and subtract it by successive interference cancellation, we introduce the constraint of $\Gamma_{12} \geq \Gamma_2$, where Γ_{12} and Γ_2 respectively represent the WU' SINR observed at the SU and the WU's signal at its own end. In addition, to guarantee the WU's quality-of-service (QoS), we have $\Gamma_2 \geq \Gamma_{2,req}$, where $\Gamma_{2,req}$ is the minimum SINR requirement of the WU.

Define EE (in bits/Joule/Hz) as the ratio of the system's throughput T_{total} to the total power consumption P_{total}. To maximize the system's EE, it needs to jointly optimize the transmission power p_1, p_2 and p_3. Accordingly, the EE maximization problem can be formulated as

$$\begin{aligned} P4.1:& \underset{p_1,p_2,p_3}{\operatorname{argmax}} \frac{T_{\text{total}}}{P_{\text{total}}}, \\ \text{s.t } (C1)&: p_1 + p_2 \leq p_{BS,max}, \\ (C2)&: p_3 \leq p_{3,max}, \\ (C3)&: 0 \leq p_1, 0 \leq p_2, 0 \leq p_3, \\ (C4)&: \Gamma_{12} \geq \Gamma_2, \ (C5): \Gamma_2 \geq \Gamma_{2,req}. \end{aligned} \tag{4.1}$$

Constraint $(C1)$ indicates that the transmission power allocated at the BS should be lower than the available transmission power budget $p_{BS,max}$. Constraint $(C2)$ denotes that the transmission power of the SU should be lower than its total available transmission power budget $p_{3,max}$. Constraint $(C3)$ implies that all the transmission power should be non-negative. Constraint $(C4)$ guarantees that the SU can decode the WU's signal, while $(C5)$ ensures that the WU's QoS should be always satisfied.

4.2 Throughput and Power Consumption Analysis

The optimization problem involves both the throughput and power consumption, which will be discussed in Sects. 4.2.1 and 4.2.2, respectively.

4.2.1 Throughput Analysis

At the WU, the received signal arrives both from the BS and SU, which is given by

$$r_2[i] = (\sqrt{p_1}x_1[i] + \sqrt{p_2}x_2[i])h_2 + \sqrt{p_3}x_2[i-\tau]h_{12} + n[i], \tag{4.2}$$

where $x_1[i]$ and $x_2[i]$ are the i-th data symbols intended for the SU and the WU, respectively. $n[i]$ is the additive White Gaussian noise. The symbol delay τ is caused by the processing delay at the SU and h_2 is the channel response from the BS to the WU, while h_{12} is the channel response from the SU to the WU. All propagation channels capture the effects of large-scale and small-scale fading. Since the phase of h_2 and h_{12} is different, the desired signal replicas arriving from the BS and the SU are mis-aligned in phase. Hence, the phase of the transmitted signal at the SU has to be shifted to align the pair of received signals. The channel spanning from the BS to the WU can be represented as $h_2 = |h_2|e^{j\theta_2}$, where $|h_2|$ and $e^{j\theta_2}$ are the magnitude and phase of the channel h_2, respectively. Similarly, the channel spanning from the SU to the WU is represented as $h_{12} = |h_{12}|e^{j\theta_{12}}$, where $|h_{12}|$ and $e^{j\theta_{12}}$

are the magnitude and phase of the channel h_{12}, respectively. Therefore, the phase difference between the two channels is calculated as $\phi = \theta_2 - \theta_{12}$. To align the two signals, the signal transmitted from the SU is pre-processed as $(\sqrt{p_3}e^{j\phi})x_2$. As a result, the signal received at the WU becomes

$$r_2[i] = \big(\sqrt{p_3}|h_{12}|e^{j(\theta_{12}+\phi)}x_2[i-\tau]+ \\ \sqrt{p_2}|h_2|e^{j\theta_2}x_2[i]\big) + \sqrt{p_1}|h_2|e^{j\theta_2}x_1[i] + n[i]. \tag{4.3}$$

As seen, the pair of desired signals arriving from the BS and the SU are now well aligned at the WU,[1] and the SINR of the WU at its own end is given by

$$\Gamma_2 = \frac{p_2|h_2|^2 + p_3|h_{12}|^2}{p_1|h_2|^2 + \sigma^2}. \tag{4.4}$$

The signal received at the SU is given by

$$r_1[i] = (\sqrt{p_1}x_1[i] + \sqrt{p_2}x_2[i])h_1 + \\ (\sqrt{p_3}e^{j(\phi+\theta_r)}x_2[i-\tau])\frac{h_r}{\sqrt{\alpha}} + n[i], \tag{4.5}$$

where h_1 and h_r represent the channel spanning from the BS to the SU and the SI leakage channel from the SU's transmitter to its receiver, respectively. Still referring to (4.5), α is the SI reduction factor defined as the ratio of the SI powers before and after SI suppression. Therefore, the WU's SINR observed at the SU is given by

$$\Gamma_{12} = \frac{p_2|h_1|^2}{p_3|\tilde{h}_r|^2 + p_1|h_1|^2 + \sigma^2}, \tag{4.6}$$

where $\tilde{h}_r$ is the residual SI channel. To guarantee successful SIC at the SU, we have to ensure that $\Gamma_{12} \geq \Gamma_2$. Based on (4.4) and (4.8), the inequality $\Gamma_{12} \geq \Gamma_2$ is further derived into

$$\frac{p_2|h_1|^2}{p_3|\tilde{h}_r|^2 + p_1|h_1|^2 + \sigma^2} \geq \frac{p_2|h_2|^2 + p_3|h_{12}|^2}{p_1|h_2|^2 + \sigma^2}. \tag{4.7}$$

[1] The signal received at the WU has a low time delay, which can be mitigated by an equalizer or sequence detector. Hereby, we assume that the signal from the BS and the SU can be readily combined by the WU [95].

Since $|h_2|^2 \leq |h_1|^2$, we have

$$\begin{aligned}\Delta &= (p_1|h_1|^2|h_{12}|^2 + |h_{12}|^2\sigma + p_2|h_2|^4)^2 \\ &- 4|h_{12}|^2|h_2|^2 p_2(|h_2|^2 - |h_1|^2)\sigma^2 \geq 0.\end{aligned} \tag{4.8}$$

As a result, the feasible domain of p_3 can be written in a closed form

$$p_3 \in [0, \frac{-B + \sqrt{\Delta}}{2A}], \tag{4.9}$$

where we have $B = p_1|h_1|^2|h_{12}|^2 + |h_{12}|^2\sigma + p_2|h_2|^4$ and $A = |h_{12}|^2|h_2|^2$. Finally, the SU's SINR after SIC is calculated as

$$\Gamma_1 = \frac{p_1|h_1|^2}{p_3|\widetilde{h}_r|+\sigma^2}. \tag{4.10}$$

4.2.2 *Power Consumption Analysis*

The total power consumption mainly consists of the PA power, circuit power of the transmit/receive chains, and the power consumed by SI cancellation [20].

(a) The PA power is closely related to the radiated transmit power and to the drain efficiency of the PA, which is given by $\frac{1}{\eta}(p_1 + p_2 + p_3)$ [20]. η is the drain efficiency of the PAs. Without loss of generality, we assume that all PAs of the BS and of the users have the same drain efficiency performance.
(b) The circuit power consumption of the transmit/receive chains is proportional to the number of active transmit and receive chains, including the power consumed by the digital-to-analog converter, by the filter and synthesizer, etc. The dynamic power consumption can be calculated as $2(p_{c,r} + p_{c,t})$, where $p_{c,t}$ and $p_{r,t}$ denote the circuit power consumed by the receive and transmit chains of the BS, of the SU and of the WU. The multiplier 2 indicates that we have two transmit chains and two receive chains in the system.
(c) For the SU acting as a FD relay, additional power p_{SI} is required for SI cancellation, which can be modeled by a constant. Generally speaking, complex SI cancellation schemes dissipate higher power by their digital-to-analog converter, transmit radio unit and adders [20].
(d) The fixed power consumption p_{fix} represents power consumed by the power supply, by the active cooling system, etc. This part is independent of the state of the transmit/receive chains [20]. Finally, the total power consumption is given by

$$P_{\text{total}} = \frac{p_1 + p_2 + p_3}{\eta} + 2(p_{c,r} + p_{c,t}) + p_{SI} + p_{fix}. \tag{4.11}$$

4.3 Energy Efficiency Performance Optimization

For ease of notation, we collect the transmission power into a vector $\boldsymbol{p} = [p_1, p_2, p_3]^T$ and write $\{\boldsymbol{\Theta}\}$ as the feasible domain confined by the constraints. Hence, the EE maximization problem is re-formulated as

$$\begin{aligned} P4.2 : \underset{\boldsymbol{p}\in\{\boldsymbol{\Theta}\}}{\operatorname{argmax}} & \frac{\log_2(1+\frac{\boldsymbol{a_1p}|h_1|^2}{\boldsymbol{a_3p}|\widetilde{h_r}|+\sigma^2})+\log_2(1+\frac{\boldsymbol{a_2p}|h_2|^2+\boldsymbol{a_3p}|h_{12}|^2}{\boldsymbol{a_1p}|h_2|^2+\sigma^2})}{\frac{1}{\eta}(||\boldsymbol{p}||_1)+2(p_{c,r}+p_{c,t})+p_{fix}+p_{SI}}, \\ \text{s.t } (\hat{C1}) &: \boldsymbol{a_1p}+\boldsymbol{a_2p} \leq p_{BS,max}, \\ (\hat{C2}) &: \boldsymbol{a_3p} \leq p_{3,max}, \\ (\hat{C3}) &: \boldsymbol{0} \preceq \boldsymbol{p}, \\ (C4) &: \Gamma_{12} \geq \Gamma_2,\ (C5) : \Gamma_2 \geq \Gamma_{2,req}, \end{aligned} \tag{4.12}$$

where $\boldsymbol{a_1} = [1, 0, 0]$, $\boldsymbol{a_2} = [0, 1, 0]$ and $\boldsymbol{a_3} = [0, 0, 1]$. The operator $||\cdot||_1$ represents the 1-norm result. It is found that the objective function is non-convex and constraints $(C4)$ $(C5)$ are given in quadratic forms. To make a balance between the performance and complexity, a low-complexity near-optimal solution is desirable. Since the numerator of the objective is the difference of two convex functions, we first reformulate the objective function of (4.12) as

$$\underset{\boldsymbol{p}\in\{\boldsymbol{\Theta}\}}{\operatorname{argmin}} \frac{f(\boldsymbol{p})-g(\boldsymbol{p})}{\frac{1}{\eta}||\boldsymbol{p}||_1+2(p_{c,r}+p_{c,t})+p_{fix}+p_{SI}}. \tag{4.13}$$

where the value of $f(\boldsymbol{p})$ and $g(\boldsymbol{p})$ are calculated as

$$\begin{aligned} f(\boldsymbol{p}) = &-\log_2\big(\boldsymbol{a_3p}|\widetilde{h_r}|^2+\boldsymbol{a_1p}|h_1|^2+\sigma^2\big) \\ &-\log_2\big(\boldsymbol{a_1p}|h_2|^2+\boldsymbol{a_2p}|h_2|^2+\boldsymbol{a_3p}|h_{12}|^2+\sigma^2\big), \end{aligned} \tag{4.14}$$

$$g(\boldsymbol{p}) = -\log_2\big(\boldsymbol{a_1p}|h_2|^2+\sigma^2\big)-\log_2\big(\boldsymbol{a_3p}|\widetilde{h_r}|^+\sigma^2\big). \tag{4.15}$$

Hence, the Frank-Wolfe method of [43] can be used, which approximates $g(\boldsymbol{p})$ by its first-order Taylor series, and iteratively updates the first-order Taylor approximation along the directions that approaches the original function. The first-order Taylor approximation of $g(\boldsymbol{p})$ at the i-th iteration is given by

$$\begin{aligned} g^{(n)}(\boldsymbol{p}) = &-\log_2\big(\boldsymbol{a_1p}^{(n)}|h_2|^2+\sigma^2\big)-\log_2\big(\boldsymbol{a_3p}^{(n)}|\widetilde{h_r}|^+\sigma^2\big)- \\ &\frac{|h_2|^2\boldsymbol{a_1}(\boldsymbol{p}-\boldsymbol{p}^{(n)})}{\ln 2(\boldsymbol{a_1p}^{(n)}|h_2|^2+\sigma^2)}-\frac{|h_r|^2\boldsymbol{a_3}(\boldsymbol{p}-\boldsymbol{p}^{(n)})}{\ln 2(\boldsymbol{a_3p}^{(n)}|h_r|^2+\sigma^2)}, \end{aligned} \tag{4.16}$$

where $\boldsymbol{p}^{(n)}$ is the value of $\boldsymbol{p}$ at the n-th iteration. Now, the overall throughput is the difference between a convex function $f(\boldsymbol{p})$ and an affine function $g^{(n)}(\boldsymbol{p})$. Let us state Theorem 1 to solve the problem.

Theorem 4.1 *The reformulated problem* $\frac{f(\boldsymbol{p})-g^{(n)}(\boldsymbol{p})}{P_{\text{total}}(\boldsymbol{p})}$, *is jointly quasi-convex with respect to the variables* $\boldsymbol{p}$ *in the feasible domain.*

The proof of Theorem 4.1 is similar to that of Theorem 2.1, which thus is omitted.

Evidently, $P_{\text{total}}(\boldsymbol{p})$ is linear with respect to the variables, while $[f(\boldsymbol{p})-g^{(n)}(\boldsymbol{p})]$ is strictly jointly convex with respect to the variables. Therefore, the summation is strictly convex with respect to the variables, and $\omega(\boldsymbol{p})$ is quasi-convex with respect to the variables in $\boldsymbol{p} \in \{\boldsymbol{\Theta}\}$. For the fractional structured quasi-convex problem of (4.12), $\beta = \frac{f(\boldsymbol{p})-g^{(n)}(\boldsymbol{p})}{P_{\text{total}}(\boldsymbol{p})}$ can be associated with a subtract structured programming formulation of $f(\boldsymbol{p})-g^{(n)}(\boldsymbol{p})-\beta P_{\text{total}}(\boldsymbol{p})$ [43]. Then, the problem is reformulated as that of solving

$$f(\boldsymbol{p})-g^{(n)}(\boldsymbol{p})-\beta P_{\text{total}}(\boldsymbol{p}), \tag{4.17}$$

with a given β. Now we handle the constraints $(C4)$ and $(C5)$. Based on (4.9), $(C4)$ is equivalent to $\boldsymbol{p}^T\boldsymbol{A}\boldsymbol{p}+\boldsymbol{a_4}\boldsymbol{p} \le 0$, where we have

$$\boldsymbol{a_4} = [\,0, (|h_2|^2-|h_1|^2)\sigma^2, |h_{12}|^2\sigma^2\,], \tag{4.18}$$

and

$$\boldsymbol{A} = \begin{bmatrix} 0 & 0 & |h_1|^2|h_{12}|^2/2 \\ 0 & 0 & |h_2|^2|h_r|^2/2 \\ |h_1|^2|h_{12}|^2/2 & |h_2|^2|h_r|^2/2 & |h_{12}|^2\sigma^2 \end{bmatrix} \tag{4.19}$$

As the matrix $\boldsymbol{A}$ is not a semi-positive definite matrix, its elements are not confined to a convex set. Hence, we need to use the concept of Schur complement [92] of $\boldsymbol{P} = \boldsymbol{p}\boldsymbol{p}^T$ to relax $(C4)$ into

$$(\hat{C4a}) : \text{Tr}(\boldsymbol{A}\boldsymbol{P}) + a_4^T\boldsymbol{p} \le 0 \tag{4.20}$$

and constraint

$$(\hat{C4b}) : \begin{bmatrix} \boldsymbol{P} & \boldsymbol{p} \\ \boldsymbol{p}^T & 1 \end{bmatrix} \succeq 0. \tag{4.21}$$

On the other hand, constraint $(C5)$ is equivalent to

$$(\hat{C5}) : \boldsymbol{a_5}\boldsymbol{p} + \Gamma_{2,req}\sigma^2 \le 0, \tag{4.22}$$

where we have $\boldsymbol{a_5} = \left[\Gamma_{2,req}\sigma^2,\ -|h_2|^2,\ -|h_{12}|^2\right]$. After a series of transformations, the problem is given as

$$P4.3 : \underset{\boldsymbol{p}\in\{\boldsymbol{\Theta}\}}{\operatorname{argmin}}\ f(\boldsymbol{p}) - g^{(n)}(\boldsymbol{p}) - \beta P_{\text{total}}(\boldsymbol{p}),$$
$$\text{s.t } (\hat{C1}) - (\hat{C3}), (\hat{C4a}), (\hat{C4b}) \text{ and } (\hat{C5}). \tag{4.23}$$

The problem $P4.3$ in (4.23) now is a standard semi-definite programming (SDP) problem with a convex set, which can be readily solved by the CVX solver. Finally, a so-called "EE oriented FD cooperative NOMA" algorithm is demonstrated for optimizing $\boldsymbol{p}$, which solves the SDP problem $P4.3$ of (4.23) in the inner layer and updates β in the outer layer. To tighten the first-order Taylor approximation $g^{(n)}(\boldsymbol{p})$, we update the value of $\boldsymbol{p}^{(n)}$ in the specific direction that approaches the original function. Since $f(\boldsymbol{p}) - g^{(n)}(\boldsymbol{p}) - \beta P_{total}(\boldsymbol{p})$ serves as the upper bound of the original problem $f(\boldsymbol{p}) - g(\boldsymbol{p}) - \beta P_{total}(\boldsymbol{p})$, the value of the upper bound is reduced iteratively until convergence, as demonstrated by Line 4 of the Algorithm. Furthermore, the value of β is squeezed by the bisection method in the outer layer according to the accuracy factor ϵ, and the optimal $\beta^\star$ can be found after a number of update. Hence, the tightness and convergence of the proposed algorithm is confirmed. The whole procedure is summarized in Algorithm 4.

Algorithm 4 EE oriented FD cooperative NOMA algorithm

Input: Left/right bounds β_l and β_r, channel condition, i.e., $h_1, h_2, h_{12}, \widetilde{h_r}$, and power consumption parameters, i.e., η, $p_{c,t}$, $p_{r,t}$, p_{SI}, p_{fix}.

Output: Optimal transmission power vector to $\boldsymbol{p}^*$.

1: Set the accuracy factor to $\epsilon > 0$, and assume that $\mathcal{F}(\beta)$ is the optimal value of $f(\boldsymbol{p}) + g^{(n)}(\boldsymbol{p}) - \beta P_{\text{total}}(\boldsymbol{p})$. Let us initialize the left bound β_l and the right bound β_r for ensuring that $\mathcal{F}(\beta_l) \cdot \mathcal{F}(\beta_r) < 0$.
2: **while** $\beta_r - \beta_l > \epsilon$ **do**
3: $\quad\beta = \frac{\beta_r+\beta_l}{2}$.
4: $\quad$Solve the problem $P3$ using the Frank-Wolfe method until convergence.
5: $\quad$**if** $\mathcal{F}(\beta_l) \cdot \mathcal{F}(\beta) < 0$ **then**
6: $\quad\quad\beta_r = \beta$.
7: $\quad$**else**
8: $\quad\quad\beta_l = \beta$.
9: $\quad$**end if**
10: **end while**

4.3.1 Computational Complexity and Properties Discussions for FD WPT Relay Systems

We now check the computational complexity of the whole algorithm, and provide some interesting discussions for FD WPT relay systems. Assume that $\beta_1 = \frac{\beta_r+\beta_l}{2}$ is the midpoint of the initial interval, and β_n is the midpoint of the interval in the n-th step. Then the difference between β_n and β^* is bounded by $|\beta_n - \beta^*| \leq \frac{\beta_r - \beta_l}{2^n}$. Given a tolerance factor ϵ, the required number of iterations is given by $n \leq log_2(\frac{\beta_r-\beta_l}{\epsilon})$. In the algorithm, the left bound β_l can be set to 0. Then the value of $\mathcal{F}(\beta_l)$ is definitely positive. Furthermore, a sufficient large value of β_r can be chosen as the right bound for making the value of $\mathcal{F}(\beta_r)$ negative. Hence, the function $\mathcal{F}(\cdot)$ has opposite signs at the two bounds and thus the classic bisection method readily leads to convergence. In the inner layer, the CVX solver proposes to use an interior-point method to solve the SDP problem, which belongs to the class of path-following methods and leads to a rapid convergence. Upon denoting the complexity of the inner layer by ξ, and the complexity order of the proposed algorithm becomes $\mathcal{O}(\log_2(\frac{\beta_r-\beta_l}{\epsilon}) \cdot \xi)$.

Power Allocation Policy for Energy Efficient WPT FD Systems: A throughput maximization design aims for achieving a high throughput regardless of the power consumption. Hence, the BS only has to satisfy the WU's SINR threshold, while allocating all the remaining power to the SU. Then the SU utilizes a high transmission power to cooperate the WU. By contrast, a power minimization design aims for reducing the total power consumption and thus merely satisfies the WU's QoS constraint by a minimum transmission power. As a better choice, EE maximization design improves the ratio between the throughput and total power consumption, striking a trade-off between the throughput maximization and power minimization problems. As demonstrated, $\omega(\boldsymbol{p})$ is quasi-convex (thus the re-formulated EE is quasi-concave) with respect to the variables. For a quasi-concave function, utilizing all transmission power may not always lead to the optimal EE solution, because the power consumption in the denominator is also increases, hence the EE may be decreased. For the purpose of illustration, we consider a simple EE maximization problem in Fig. 4.2, where the EE curve is a quasi-concave function of transmission power. Neither the power minimization (point (a)) nor the throughput maximization solution (point (c)) achieves the optimal EE.

The Impact of Communication Distance on the Relaying Strategy: The distances of BS-SU, BS-WU and SU-WU also affect the SIC failure region. With a shorter BS-SU distance, the value of Γ_{12} is increased due to the reduced PL, and thus a higher cooperative power may be assigned to the SU hence guaranteeing $\Gamma_{12} \geq \Gamma_2$. A similar trend can be observed for a shorter BS-WU distance. By contrast, a longer SU- WU distance makes $\Gamma_{12} \geq \Gamma_2$ more likely. However, this also makes the cooperation between the SU-WU less energy efficient, since additional power is consumed at the SU without an obvious throughput enhancement. As expected, when the SU-WU distance tends to infinity, the cooperative power assigned to the SU tends to 0, results in non-cooperative NOMA systems.

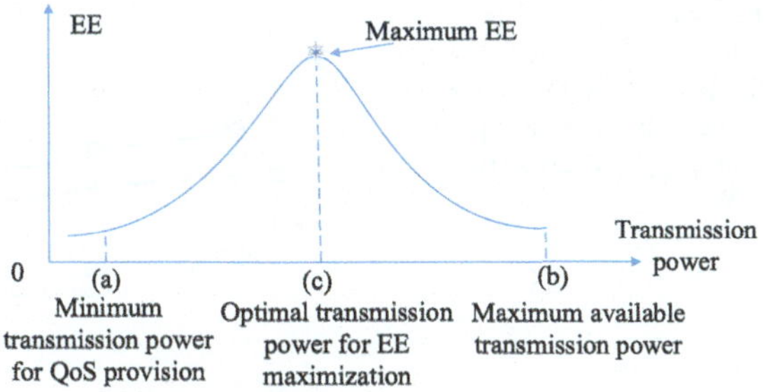

Fig. 4.2 The stylized relationship of EE maximization, power minimization and throughput maximization problems

Table 4.1 Simulation setup of FD cooperative NOMA systems

Bandwidth	1.25 MHz
AWGN power spectral density	−174 dBm/Hz
Drain efficiency of PA η	35%
$p_{c,r}$ and $p_{c,t}$	100 mW
p_{SI} and p_{fix}	50 and 500 mW
$p_{BS,max}$ and $p_{3,max}$	100 and 20 mW
SI cancellation amount	80 dB
Distance between the BS and the WU	200 m

The Impact of Drain Effieicny on the Relaying Strategy: With a lower drain efficiency, the system's EE is reduced due to the boosted PA power consumption. Also, since the throughput improvement attained by increasing the transmission power may not be in line with the increased power consumption, the power allocation strategy of both the BS and of the SU becomes more conservative.

4.4 Performance Analysis

The simulation parameters are given in Table 4.1. The 3GPP standards operating at 2 GHz PL model of $\text{PL}(d) = 145.4 + 37.5\log_{10}(d/1000)$ [43] and log normal shadowing model are used. The small-scale fading is modeled by Rayleigh fading except for the SI channel at the SU, which is modeled as Rician fading having a Rician factor of 5 dB [95]. A pair of classic cooperative NOMA systems are selected for performance comparison. (a) In HD cooperative NOMA systems [96], the BS transmits its downlink signals to the two users (with transmission power $p_1 = \alpha p_{BS,max}$, $p_2 = (1-\alpha)p_{BS,max}$, and $\alpha = 0.2$) during the first half of the time slot, while the SU relays the WU's signal with its full transmission power $p_3 = p_{3,max}$ in the second half of the time slot. (b) In the pre-fixed FD cooperative NOMA systems [95], the SU acts as a FD relay node for simultaneously helping the WU by assigning its full transmission power $p_3 = p_{3,max}$, whereas the power allocation at the BS is fixed as $p_1 = \alpha p_{BS,max}$, $p_2 = (1-\alpha)p_{BS,max}$, and $\alpha = 0.2$.

Figure 4.3a, b demonstrate the impact of the normalized distance between the BS and the SU on the EE and SE performance. The algorithm can outperform other

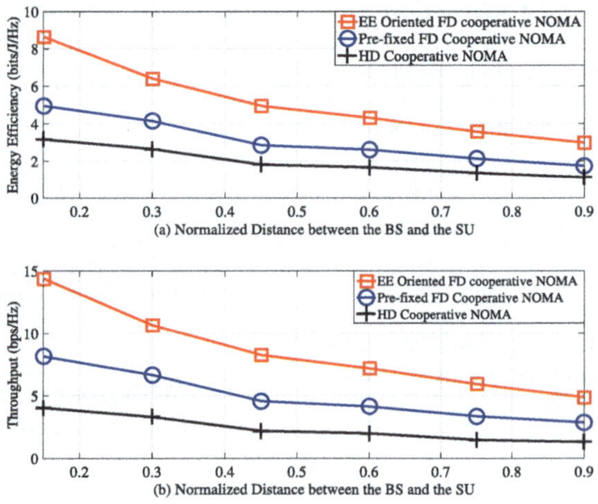

Fig. 4.3 Impact of the normalized distance between the BS and the SU on the value of EE and throughput, with $p_{3,max} = 20 \times 10^{-3}$ mW

comparison algorithms in terms of its EE, whilst exhibiting a higher robustness against the SU's location. This is because the benchmarking algorithms consume all the available transmission power, which degrades their EE. A further degrading factor is constituted by their low throughput. In Fig. 4.3b, the designed algorithm shows the highest throughput among the three algorithms. This is because the algorithm enables an effective cooperation of the SU and of the WU, and thus can improve the WU's throughput. More importantly, no additional time slot is needed for the cooperation phase, and thus the proposed algorithm significantly outperforms the HD cooperative NOMA. As for the pre-fixed FD cooperative NOMA, the power allocation is independent to the relative distance among the communication nodes. Furthermore, the SU uses its full transmission power to cooperate with the WU, which may result in the "SINR gap reversal" problem and may lead an unsuccessful SIC operation at the SU side. As a result, the throughput of the pre-fixed FD cooperative NOMA is also inferior to the designed algorithm. Finally, the throughput of all the algorithms degrades for a longer distance between the BS and the SU due to the high PL.

Figure 4.4 shows the impact of the constraint $p_{3,max}$ on the SIC's failure probability, which is defined as the ratio of the times of unsuccessful SIC operation at the SU to the times of simulations. As seen, the proposed algorithm is shown to guarantee that the SU can successfully decode the WU's signal by SIC. This is because an additional constraint is imposed on the power control at the SU by the proposed algorithm. By comparison, the SIC failure probability of the pre-fixed FD cooperative NOMA systems increases rapidly with the transmission power constraint $p_{3,max}$. This is because for a higher transmission power used at the SU, the WU's SINR Γ_2 increases, while the WU's SINR Γ_{12} observed at the SU decreases due to the higher level of residual SI. As a result, a higher value of $p_{3,max}$ leads to a higher SIC failure probability at the SU. In case of poor SI cancellation, the SIC failure probability is substantially increased by the strong residual SI. As for

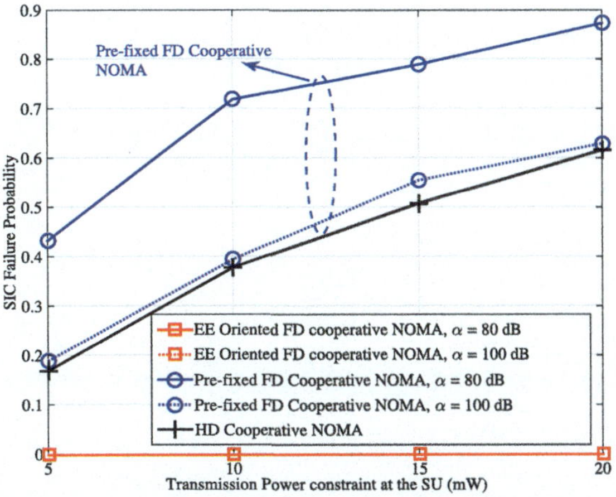

Fig. 4.4 Impact of the maximum constraint $p_{3,max}$ at the SU on the SIC's failure probability, where the SU is in the middle of the BS and the WU

the HD cooperative NOMA system, its SIC failure probability also increases upon imposing a higher transmission power constraint due to the "SINR gap reversal" problem. Since no residual SI is imposed by the HD operation, its SIC failure probability is lower than that of the FD operation [95]. However, its SIC failure probability still remains significantly higher than that of the proposed algorithm, because the HD cooperative NOMA neglects that Γ_2 may be higher than Γ_{12} owing to the cooperation between the SU and WU. However, its SIC failure probability is not affected by the SI cancellation factor α in the presence of HD operation.

Appendix A
Proof of Theorem 2.1

Define the superlevel set of $\omega(\boldsymbol{W}, \boldsymbol{p})$ as $S_\delta = \{\boldsymbol{W}, \boldsymbol{p} \in \boldsymbol{\Theta} | \omega(\boldsymbol{W}, \boldsymbol{p}) \geq \delta\}$. According to [26], $\omega(\boldsymbol{W}, \boldsymbol{p})$ is jointly quasi-concave with respect to variables $\boldsymbol{W}$ and $\boldsymbol{p}$ if S_δ is convex for any real number δ. When $\delta \leq 0$, there is no physical meaning. When $\delta \geq 0$, S_δ is equivalent to $S_\delta = \{\boldsymbol{W}, \boldsymbol{p} \in \boldsymbol{\Theta} | P_{\text{total}}(\boldsymbol{W}, \boldsymbol{p}) - \delta\Big(f_1(\boldsymbol{W}, \boldsymbol{p}) - f_2^{(n)}(\boldsymbol{W}, \boldsymbol{p})\Big) \leq 0\}$. According to our analysis, $P_{\text{total}}(\boldsymbol{W}, \boldsymbol{p})$ is affine with respect to the variables, while $-\delta\Big(f_1(\boldsymbol{W}, \boldsymbol{p}) - f_2^{(n)}(\boldsymbol{W}, \boldsymbol{p})\Big)$ is strictly jointly convex with respect to the variables. Therefore, the summation is strictly convex with respect to the variables, and $\omega(\boldsymbol{W}, \boldsymbol{p})$ is quasi-concave with respect to the variables.

Z. Wei et al., *Energy Efficient Full Duplex Wireless Communication Systems*, Wireless Networks, https://doi.org/10.1007/978-981-97-8915-3

Appendix B
Equivalence Between Maximizing Problem $P2.4$ and Finding the Root of Problem $P2.5$

The proof is similar to and based on the Lemma 1 in [44]. Suppose that $(\boldsymbol{W}^*, \boldsymbol{p}^*)$ is the optimal solution of (4.23), with the optimal β^*. Define β_0 as the root of $f_1(\boldsymbol{W}, \boldsymbol{p}) - f_2^{(n)}(\boldsymbol{W}, \boldsymbol{p}) - \beta P_{\text{total}}(\boldsymbol{W}, \boldsymbol{p}) = 0$. If $f_1(\boldsymbol{W}, \boldsymbol{p}) - f_2^{(n)}(\boldsymbol{W}, \boldsymbol{p}) - \beta_0 P_{\text{total}}(\boldsymbol{W}, \boldsymbol{p}) = 0$ is obtained at $(\boldsymbol{W_0}, \boldsymbol{p_0})$, we know that there is at least one solution $(\boldsymbol{W_0}, \boldsymbol{p_0})$ satisfying $f_1(\boldsymbol{W_0}, \boldsymbol{p_0}) - f_2^{(n)}(\boldsymbol{W_0}, \boldsymbol{p_0}) - \beta_0 P_{\text{total}}(\boldsymbol{W_0}, \boldsymbol{p_0}) = 0$. Also, we have

$$\beta^* \geq \frac{f_1(\boldsymbol{W_0}, \boldsymbol{p_0}) - f_2^{(n)}(\boldsymbol{W_0}, \boldsymbol{p_0})}{P_{\text{total}}(\boldsymbol{W_0}, \boldsymbol{p_0})} = \beta_0. \tag{B.1}$$

Due to the optimality, $f_1(\boldsymbol{W}, \boldsymbol{p}) - f_2^{(n)}(\boldsymbol{W}, \boldsymbol{p}) - \beta_0 P_{\text{total}}(\boldsymbol{W}, \boldsymbol{p}) = 0$ means $f_1(\boldsymbol{W}, \boldsymbol{p}) - f_2^{(n)}(\boldsymbol{W}, \boldsymbol{p}) - \beta_0 P_{\text{total}}(\boldsymbol{W}, \boldsymbol{p}) \leq 0$ for all $\boldsymbol{W}, \boldsymbol{p} \in \{\boldsymbol{\Theta}\}$, including $\boldsymbol{W}^*, \boldsymbol{p}^*$. Therefore, we have $\beta^* \leq \beta_0$. As a result, $\beta_* = \beta_0$ is led.

On the other hand, if $\beta = \beta^*$, at least we have $f_1(\boldsymbol{W}^*, \boldsymbol{p}^*) - f_2^{(n)}(\boldsymbol{W}^*, \boldsymbol{p}^*) - \beta P_{\text{total}}(\boldsymbol{W}^*, \boldsymbol{p}^*) = 0$ and thus $f_1(\boldsymbol{W}, \boldsymbol{p}) - f_2^{(n)}(\boldsymbol{W}, \boldsymbol{p}) - \beta^* P_{\text{total}}(\boldsymbol{W}, \boldsymbol{p}) \geq 0$. Also, because β^* is optimal from (D.1), for all feasible $(\boldsymbol{W}, \boldsymbol{p}) \in \{\boldsymbol{\Theta}\}$, $f_1(\boldsymbol{W}, \boldsymbol{p}) - f_2^{(n)}(\boldsymbol{W}, \boldsymbol{p}) - \beta^* P_{\text{total}}(\boldsymbol{W}, \boldsymbol{p}) \leq 0$ holds. As a result, the equality $\beta = \beta^*$ yields $f_1(\boldsymbol{W}, \boldsymbol{p}) - f_2^{(n)}(\boldsymbol{W}, \boldsymbol{p}) - \beta^* P_{\text{total}}(\boldsymbol{W}, \boldsymbol{p}) = 0$. In conclusion, the sufficient and necessary condition of $\beta = \beta^*$ is $f_1(\boldsymbol{W}, \boldsymbol{p}) - f_2^{(n)}(\boldsymbol{W}, \boldsymbol{p}) - \beta P_{\text{total}}(\boldsymbol{W}, \boldsymbol{p}) = 0$, and the equivalence between maximizing problem $P2.4$ and finding the root of $P2.5$ is proven.

Z. Wei et al., *Energy Efficient Full Duplex Wireless Communication Systems*,
Wireless Networks, https://doi.org/10.1007/978-981-97-8915-3

Appendix C
Proof of Rank One

The SDP relaxed problem in (4.23) satisfies the Slater's constraint qualification. To obtain the dual problem, the Lagrange function of the primal problem is given in (E.4). According to (E.4), the KKT conditions used for the proof are given by (C.1)

$$\begin{aligned}
&\frac{K\sum_{m=1}^{L}\sum_{n=1}^{L} t_m t_n \boldsymbol{F}_{\boldsymbol{m}}^T(\boldsymbol{F}_{\boldsymbol{n}}^H)^T \boldsymbol{H}_k}{\Gamma^*} - \sum_{k'\neq K}^{K} \frac{\sum_{m=1}^{L}\sum_{n=1}^{L} t_m t_n \boldsymbol{F}_{\boldsymbol{m}}^T(\boldsymbol{F}_{\boldsymbol{n}}^H)^T \boldsymbol{H}_{\boldsymbol{k'}}}{a_{k'}^{(n)}} + \\
&\frac{U\mathrm{Tr}\big(\mathrm{diag}(\boldsymbol{h}_{\boldsymbol{SI}}(\sum_{k=1}^{K}\sum_{m=1}^{L}\sum_{n=1}^{L} t_m t_n \boldsymbol{F}_{\boldsymbol{m}}^T(\boldsymbol{F}_{\boldsymbol{n}}^H))^T \boldsymbol{h}_{\boldsymbol{SI}}^H) \quad (\sum_{i=1}^{L}\sum_{j=1}^{L} r_i r_j \boldsymbol{F}_{\boldsymbol{i}}\boldsymbol{G}_{\boldsymbol{u}}\boldsymbol{F}_{\boldsymbol{j}}^H)\big)}{\alpha\Lambda^*} \\
&-\sum_{u=1}^{U} \frac{\mathrm{diag}(\boldsymbol{h}_{\boldsymbol{SI}}(\sum_{k=1}^{K}\sum_{m=1}^{L}\sum_{n=1}^{L} t_m t_n \boldsymbol{F}_{\boldsymbol{m}}^T(\boldsymbol{F}_{\boldsymbol{n}}^H))^T \boldsymbol{h}_{\boldsymbol{SI}}^H) \quad (\sum_{i=1}^{L}\sum_{j=1}^{L} r_i r_j \boldsymbol{F}_{\boldsymbol{i}}\boldsymbol{G}_{\boldsymbol{u}}\boldsymbol{F}_{\boldsymbol{j}}^H)}{\alpha b_u^{(n)}} \\
&-\frac{\beta}{\eta}\boldsymbol{I} - \sum_{l=1}^{L} \zeta_l^* \boldsymbol{F}_{\boldsymbol{l}}^T + \boldsymbol{Z}_{\boldsymbol{k}}^* = 0;\ \boldsymbol{W}_{\boldsymbol{k}}^*\boldsymbol{Z}_{\boldsymbol{k}}^* = \boldsymbol{0};\ \boldsymbol{Z}_{\boldsymbol{k}}^* \succeq \boldsymbol{0};\ \zeta_l^* \geq 0;
\end{aligned} \tag{C.1}$$

where Γ^* are Λ^* are both scalars, such that

$$\Gamma^* = \sum_{k=1}^{K} \log_2\Big(\sum_{k'=1}^{K} \mathrm{Tr}(\sum_{m=1}^{L}\sum_{n=1}^{L} t_m t_n \boldsymbol{F}_{\boldsymbol{m}} \boldsymbol{W}_{\boldsymbol{k'}}{}^* \boldsymbol{F}_{\boldsymbol{n}}^H \boldsymbol{H}_{\boldsymbol{k}}) + \sum_{u=1}^{U} p_u^* |e_{u,k}|^2 + \sigma_k^2\Big), \tag{C.2}$$

Z. Wei et al., *Energy Efficient Full Duplex Wireless Communication Systems*, Wireless Networks, https://doi.org/10.1007/978-981-97-8915-3

and

$$\Lambda^* = \sum_{u=1}^{U} \log_2\Big(\sum_{u' \neq u}^{U} p_{u'}^* \mathrm{Tr}(\boldsymbol{G}_{u'} \sum_{i=1}^{L}\sum_{j=1}^{L} r_i r_j \boldsymbol{F_i} \boldsymbol{G_u} \boldsymbol{F}_j^H) +$$
$$\sigma^2 \mathrm{Tr}(\sum_{i=1}^{L}\sum_{j=1}^{L} r_i r_j \boldsymbol{F_i} \boldsymbol{G_u} \boldsymbol{F}_j^H) +$$
$$\frac{1}{\alpha} \mathrm{Tr}\big(\mathrm{diag}(\boldsymbol{h_{SI}}(\sum_{k=1}^{K}\sum_{m=1}^{L}\sum_{n=1}^{L} t_m t_n \boldsymbol{F_m} \boldsymbol{W_k}^* \boldsymbol{F}_n^H) \boldsymbol{h}_{SI}^H)(\sum_{i=1}^{L}\sum_{j=1}^{L} r_i r_j \boldsymbol{F_i} \boldsymbol{G_u} \boldsymbol{F}_j^H)\big)\Big) \tag{C.3}$$

For simplicity, the Eq. (C.1) can be re-expressed by (C.4). By premultiplying both side of the first equality in (C.4) by $\boldsymbol{W}_k^*$, we obtain

$$\boldsymbol{Z_k}^* = \boldsymbol{A_k}^* - \frac{K \sum_{m=1}^{L} \sum_{n=1}^{L} t_m t_n \boldsymbol{F}_m^T (\boldsymbol{F}_n^H)^T \boldsymbol{H_k}}{\Gamma^*},$$
$$\boldsymbol{A_k}^* = \sum_{k' \neq K}^{K} \frac{\sum_{m=1}^{L} \sum_{n=1}^{L} t_m t_n \boldsymbol{F}_m^T (\boldsymbol{F}_n^H)^T \boldsymbol{H_{k'}}}{a_{k'}^{(n)}} + \frac{\beta}{\eta} \boldsymbol{I} + \sum_{l=1}^{L} \varsigma_l^* \boldsymbol{F}_l^T +$$
$$\sum_{u=1}^{U} \frac{\mathrm{diag}(\boldsymbol{h_{SI}}(\sum_{k=1}^{K}\sum_{m=1}^{L}\sum_{n=1}^{L} t_m t_n \boldsymbol{F}_m^T (\boldsymbol{F}_n^H))^T \boldsymbol{h}_{SI}^H)(\sum_{i=1}^{L}\sum_{j=1}^{L} r_i r_j \boldsymbol{F_i} \boldsymbol{G_u} \boldsymbol{F}_j^H))}{\alpha b_u^{(n)}} \tag{C.4}$$
$$- \frac{U \mathrm{Tr}\big(\mathrm{diag}(\boldsymbol{h_{SI}}(\sum_{k=1}^{K}\sum_{m=1}^{L}\sum_{n=1}^{L} t_m t_n \boldsymbol{F}_m^T (\boldsymbol{F}_n^H))^T \boldsymbol{h}_{SI}^H)(\sum_{i=1}^{L}\sum_{j=1}^{L} r_i r_j \boldsymbol{F_i} \boldsymbol{G_u} \boldsymbol{F}_j^H))}{\alpha \Lambda^*}$$

$$\boldsymbol{W}_k^* \boldsymbol{A_k}^* = \boldsymbol{W}_k^* \frac{K \sum_{m=1}^{L} \sum_{n=1}^{L} t_m t_n \boldsymbol{F}_m^T (\boldsymbol{F}_n^H)^T \boldsymbol{H_k}}{\Gamma}. \tag{C.5}$$

Since all channel variables in the considered system are statistically independent, the matrices span the whole signal space, and it can be known that $\boldsymbol{A_k}^*$ is a full-rank matrix [45]. According to the properties of matrix calculus, we have

$$
\begin{aligned}
\mathrm{Rank}(\boldsymbol{W}_{\boldsymbol{k}}^{*}) &= \mathrm{Rank}(\boldsymbol{W}_{\boldsymbol{k}}^{*}\boldsymbol{A}_{k}) \\
&= \mathrm{Rank}(\boldsymbol{W}_{\boldsymbol{k}}^{*}\frac{K\sum_{m=1}^{L}\sum_{n=1}^{L}t_m t_n \boldsymbol{F}_{\boldsymbol{m}}^{T}(\boldsymbol{F}_{\boldsymbol{n}}^{H})^{T}\boldsymbol{H}_{k}}{\Gamma}) \\
&\le \min\{\mathrm{Rank}(\boldsymbol{W}_{\boldsymbol{k}}^{*}), \mathrm{Rank}(\frac{K\sum_{m=1}^{L}\sum_{n=1}^{L}t_m t_n \boldsymbol{F}_{\boldsymbol{m}}^{T}(\boldsymbol{F}_{\boldsymbol{n}}^{H})^{T}\boldsymbol{H}_{k}}{\Gamma})\} \\
&\le \mathrm{Rank}(\frac{K\sum_{m=1}^{L}\sum_{n=1}^{L}t_m t_n \boldsymbol{F}_{\boldsymbol{m}}^{T}(\boldsymbol{F}_{\boldsymbol{n}}^{H})^{T}\boldsymbol{H}_{k}}{\Gamma}).
\end{aligned}
\tag{C.6}
$$

Since $\mathrm{Rank}(\frac{K\sum_{m=1}^{L}\sum_{n=1}^{L}t_m t_n \boldsymbol{F}_{\boldsymbol{m}}^{T}(\boldsymbol{F}_{\boldsymbol{n}}^{H})^{T}\boldsymbol{H}_{k}}{\Gamma}) \le 1$, it is derived that the rank of matrix $\boldsymbol{W}_{\boldsymbol{k}}^{*}$ is smaller or equal to 1. It is well known that only a null matrix has rank 0 while $\boldsymbol{W}_{\boldsymbol{k}}^{*} \neq \boldsymbol{0}$. Thus, $\mathrm{Rank}(\boldsymbol{W}_{\boldsymbol{k}}^{*}) = 1$ is obtained.

Appendix D
Proof of Convergence in Outer Layer

Suppose that $(\boldsymbol{W}^*, \boldsymbol{p}^*)$ is the optimal solution, with the optimal β^*. Define β_0 as the root of $f_1(\boldsymbol{W}, \boldsymbol{p}) - f_2^{(n)}(\boldsymbol{W}, \boldsymbol{p}) - \beta p_{\text{total}}(\boldsymbol{W}, \boldsymbol{p}) = 0$.

If $f_1(\boldsymbol{W}, \boldsymbol{p}) - f_2^{(n)}(\boldsymbol{W}, \boldsymbol{p}) - \beta_0 p_{\text{total}}(\boldsymbol{W}, \boldsymbol{p}) = 0$ is obtained at $(\boldsymbol{W_0}, \boldsymbol{p_0}) \in \boldsymbol{\Theta}$, it can be known that there is at least one solution $(\boldsymbol{W_0}, \boldsymbol{p_0})$ satisfying $f_1(\boldsymbol{W_0}, \boldsymbol{p_0}) - f_2^{(n)}(\boldsymbol{W_0}, \boldsymbol{p_0}) - \beta_0 p_{\text{total}}(\boldsymbol{W_0}, \boldsymbol{p_0}) = 0$. Also, we have

$$\beta^* \geq \frac{f_1(\boldsymbol{W_0}, \boldsymbol{p_0}) - f_2^{(n)}(\boldsymbol{W_0}, \boldsymbol{p_0})}{p_{\text{total}}(\boldsymbol{W_0}, \boldsymbol{p_0})} = \beta_0. \tag{D.1}$$

Due to the optimality, $f_1(\boldsymbol{W}, \boldsymbol{p}) - f_2^{(n)}(\boldsymbol{W}, \boldsymbol{p}) - \beta_0 p_{\text{total}}(\boldsymbol{W}, \boldsymbol{p}) = 0$ means $f_1(\boldsymbol{W}, \boldsymbol{p}) - f_2^{(n)}(\boldsymbol{W}, \boldsymbol{p}) - \beta_0 p_{\text{total}}(\boldsymbol{W}, \boldsymbol{p}) \leq 0$ for all $\boldsymbol{W}, \boldsymbol{p} \in \boldsymbol{\Theta}$, including $\boldsymbol{W}^*, \boldsymbol{p}^*$. Therefore, we have $\beta^* \leq \beta_0$. As a result, $\beta_* = \beta_0$ is readily obtained.

On the other hand, if $\beta = \beta^*$, at least we have

$$f_1(\boldsymbol{W}^*, \boldsymbol{p}^*) - f_2^{(n)}(\boldsymbol{W}^*, \boldsymbol{p}^*) - \beta p_{\text{total}}(\boldsymbol{W}^*, \boldsymbol{p}^*) = 0. \tag{D.2}$$

Thus, we have

$$f_1(\boldsymbol{W}, \boldsymbol{p}) - f_2^{(n)}(\boldsymbol{W}, \boldsymbol{p}) - \beta^* p_{\text{total}}(\boldsymbol{W}, \boldsymbol{p}) \geq 0. \tag{D.3}$$

Also, because β^* is optimal, for all feasible results $(\boldsymbol{W}, \boldsymbol{p}) \in \boldsymbol{\Theta}$, $f_1(\boldsymbol{W}, \boldsymbol{p}) - f_2^{(n)}(\boldsymbol{W}, \boldsymbol{p}) - \beta^* p_{\text{total}}(\boldsymbol{W}, \boldsymbol{p}) \leq 0$ holds. As a result, the equality $\beta = \beta^*$ derives that

$$f_1(\boldsymbol{W}, \boldsymbol{p}) - f_2^{(n)}(\boldsymbol{W}, \boldsymbol{p}) - \beta^* p_{\text{total}}(\boldsymbol{W}, \boldsymbol{p}) = 0. \tag{D.4}$$

In a nutshell, the sufficient and necessary condition of $\beta = \beta^*$ is given as $f_1(\boldsymbol{W}, \boldsymbol{p}) - f_2^{(n)}(\boldsymbol{W}, \boldsymbol{p}) - \beta p_{\text{total}}(\boldsymbol{W}, \boldsymbol{p}) = 0$.

Z. Wei et al., *Energy Efficient Full Duplex Wireless Communication Systems*, Wireless Networks, https://doi.org/10.1007/978-981-97-8915-3

Appendix E
Proof of Convergence in Inner Layer

The proof of convergence of FW method follows the similar approach in [24]. Let $\Upsilon^{(n+1)}$ denote the optimal value of at the $n+1th$ iteration. According to the updating rule in FW, we have

$$\begin{aligned}
&\Upsilon^{(n+1)} = \\
&f_1(\boldsymbol{W}^{(n+1)}, \boldsymbol{p}^{(n+1)}) - f_2^{(n)}(\boldsymbol{W}^{(n+1)}, \boldsymbol{p}^{(n+1)}) - \beta P_{\text{total}}(\boldsymbol{W}^{(n+1)}, \boldsymbol{p}^{(n+1)}) \\
&\quad = \underset{\boldsymbol{W}, \boldsymbol{p} \in \Theta.}{\text{argmax}}\, f_1(\boldsymbol{W}, \boldsymbol{p}) - f_2^{(n)}(\boldsymbol{W}, \boldsymbol{p}) - \beta P_{\text{total}}(\boldsymbol{W}, \boldsymbol{p}).
\end{aligned} \tag{E.1}$$

Since $\Upsilon^{(n+1)}$ is the optimal at the $(n+1)$-th iteration, it should have a greater value than any other solution in the feasible domain. Therefore we have the following inequality

$$\begin{aligned}
&\underset{\boldsymbol{W}, \boldsymbol{p} \in \Theta.}{\text{argmax}}\, f_1(\boldsymbol{W}, \boldsymbol{p}) - f_2^{(n)}(\boldsymbol{W}, \boldsymbol{p}) - \beta P_{\text{total}}(\boldsymbol{W}, \boldsymbol{p}) \geq \\
&f_1(\boldsymbol{W}^{(n)}, \boldsymbol{p}^{(n)}) - f_2^{(n)}(\boldsymbol{W}^{(n)}, \boldsymbol{p}^{(n)}) - \beta P_{\text{total}}(\boldsymbol{W}^{(n)}, \boldsymbol{p}^{(n)})
\end{aligned} \tag{E.2}$$

Since we have the equalities that $f_2^{(n)}(\boldsymbol{W}^{(n)}, \boldsymbol{p}^{(n)}) = f_2(\boldsymbol{W}^{(n)}, \boldsymbol{p}^{(n)})$ and $\nabla f_2^{(n)}(\boldsymbol{W}^{(n)}, \boldsymbol{p}^{(n)}) = \nabla f_2(\boldsymbol{W}^{(n)}, \boldsymbol{p}^{(n)})$. Based on the two equalities above, (E.2) can be further derived as

$$\begin{aligned}
&f_1(\boldsymbol{W}^{(n)}, \boldsymbol{p}^{(n)}) - f_2^{(n)}(\boldsymbol{W}^{(n)}, \boldsymbol{p}^{(n)}) - \beta P_{\text{total}}(\boldsymbol{W}^{(n)}, \boldsymbol{p}^{(n)}) \\
&= f_1(\boldsymbol{W}^{(n)}, \boldsymbol{p}^{(n)}) - f_2(\boldsymbol{W}^{(n)}, \boldsymbol{p}^{(n)}) - \beta P_{\text{total}}(\boldsymbol{W}^{(n)}, \boldsymbol{p}^{(n)}) \\
&\geq f_1(\boldsymbol{W}^{(n)}, \boldsymbol{p}^{(n)}) - f_2^{(n-1)}(\boldsymbol{W}^{(n)}, \boldsymbol{p}^{(n)}) - \beta P_{\text{total}}(\boldsymbol{W}^{(n)}, \boldsymbol{p}^{(n)}) = \Upsilon^{(n)}.
\end{aligned} \tag{E.3}$$

Z. Wei et al., *Energy Efficient Full Duplex Wireless Communication Systems*,
Wireless Networks, https://doi.org/10.1007/978-981-97-8915-3

As a result, it indicates that $\Upsilon^{(n+1)} \geq \Upsilon^{(n)}$, meaning the value of (2.23) is non-decreasing after each iteration. Also, the value of (2.23) is upper bounded by the transmission power constraints $(\widetilde{C1})$ and $(C2)$, and thus the FW method confirms the convergence.

Then it is proven that the convergence lead to a KKT point. The Lagrange function of (2.23) is established as

$$\begin{aligned} L = f_1(\boldsymbol{W}, \boldsymbol{p}) - f_2^{(n)}(\boldsymbol{W}, \boldsymbol{p}) - \beta P_{\text{total}}(\boldsymbol{W}, \boldsymbol{p}) + \lambda_u p_u + \\ \sum_{l=1}^{L} \zeta_l (p_{DA} - \sum_{k=1}^{K} Tr(\boldsymbol{W}_k \boldsymbol{F}_l)) + \boldsymbol{W}_k \boldsymbol{Z}_k + \varsigma_u (p_{u,max} - p_u). \end{aligned} \tag{E.4}$$

The KKT condition of the optimal value at iteration n is given by the following equalities

$$\begin{aligned} &\nabla_{\boldsymbol{W}_k} f_1(\boldsymbol{W}^{(n)}, \boldsymbol{p}^{(n)}) - \nabla_{\boldsymbol{W}_k} f_2^{(n)}(\boldsymbol{W}^{(n)}, \boldsymbol{p}^{(n)}) + \boldsymbol{Z}_k - \nabla_{\boldsymbol{W}_k} \beta P_{\text{total}}(\boldsymbol{W}^{(n)}, \boldsymbol{p}^{(n)}) - \\ &\sum_{l=1}^{L} \zeta_l \boldsymbol{F}_l = 0, \\ &\nabla_{p_u} f_1(\boldsymbol{W}^{(n)}, \boldsymbol{p}^{(n)}) - \nabla_{p_u} f_2^{(n)}(\boldsymbol{W}^{(n)}, \boldsymbol{p}^{(n)}) - \varsigma_u \\ &\quad + \lambda_u - \nabla_{p_u} \beta P_{\text{total}}(\boldsymbol{W}^{(n)}, \boldsymbol{p}^{(n)}) = 0, \\ &\zeta_l (p_{DA} - \sum_{k=1}^{K} Tr(\boldsymbol{W}_k^{(n)} \boldsymbol{F}_l)) = 0, Tr(\boldsymbol{W}_{k^{(n)}} Z_k) = 0, \\ &\varsigma_u (p_{u,max} - p_u^{(n)}) = 0, \lambda_u p_u^{(n)} = 0, \end{aligned} \tag{E.5}$$

for $\forall k \in K, \forall u \in U$, and $\forall l \in L$. Based on the equality that $\nabla f_2^{(n)}(\boldsymbol{W}^{(n)}, \boldsymbol{p}^{(n)}) = \nabla f_2(\boldsymbol{W}^{(n)}, \boldsymbol{p}^{(n)})$, $\nabla_{\boldsymbol{W}_k} f_2^{(n)}(\boldsymbol{W}^{(n)}, \boldsymbol{p}^{(n)})$ and $\nabla_{p_u} f_2^{(n)}(\boldsymbol{W}^{(n)}, \boldsymbol{p}^{(n)})$ can be replaced by

$\nabla_{\boldsymbol{W}_k} f_2(\boldsymbol{W}^{(n)}, \boldsymbol{p}^{(n)})$ and $\nabla_{p_u} f_2(\boldsymbol{W}^{(n)}, \boldsymbol{p}^{(n)})$, which is exactly the KKT condition of the original problem $f_1(\boldsymbol{W}, \boldsymbol{p}) - f_2(\boldsymbol{W}, \boldsymbol{p}) - P_{\text{total}}(\boldsymbol{W}, \boldsymbol{p})$.

Appendix F
Proof of Theorem 3.1

T_1 and T_2 can be calculated as

$$T_1 = \frac{T \sum_{k=1}^{K} \sum_{n=1}^{N} \beta_{k,n} log_2(1 + \frac{p_{r,k,n}\gamma_{rd,k,n}}{\beta_{k,n}})}{\sum_{k=1}^{K} \sum_{n=1}^{N} \beta_{k,n} log_2(1 + \frac{p_{s,k,n}\gamma_{sr,k,n}}{\beta_{k,n}}) + \sum_{k=1}^{K} \sum_{n=1}^{N} \beta_{k,n} log_2(1 + \frac{p_{r,k,n}\gamma_{rd,k,n}}{\beta_{k,n}})} \tag{F.1}$$

$$T_2 = \frac{T \sum_{k=1}^{K} \sum_{n=1}^{N} \beta_{k,n} log_2(1 + \frac{p_{s,k,n}\gamma_{sr,k,n}}{\beta_{k,n}})}{\sum_{k=1}^{K} \sum_{n=1}^{N} \beta_{k,n} log_2(1 + \frac{p_{s,k,n}\gamma_{sr,k,n}}{\beta_{k,n}}) + \sum_{k=1}^{K} \sum_{n=1}^{N} \beta_{k,n} log_2(1 + \frac{p_{r,k,n}\gamma_{rd,k,n}}{\beta_{k,n}})} \tag{F.2}$$

Substituting (F.1) and (F.2) into (3.15), an equivalent object function is obtained as

$$\underset{T_1, T_2, \mathbf{p}_s, \mathbf{p}_r, \boldsymbol{\beta}}{\text{argmax}} \frac{2\left(\sum_{k=1}^{K} \sum_{n=1}^{N} \beta_{k,n} log_2(1 + \frac{p_{s,k,n}\gamma_{sr,k,n}}{\beta_{k,n}})\right)\left(\sum_{k=1}^{K} \sum_{n=1}^{N} \beta_{k,n} log_2(1 + \frac{p_{r,k,n}\gamma_{rd,k,n}}{\beta_{k,n}})\right)}{\sum_{k=1}^{K} \sum_{n=1}^{N} \beta_{k,n} log_2(1 + \frac{p_{s,k,n}\gamma_{sr,k,n}}{\beta_{k,n}}) + \sum_{k=1}^{K} \sum_{n=1}^{N} \beta_{k,n} log_2(1 + \frac{p_{r,k,n}\gamma_{rd,k,n}}{\beta_{k,n}})} \tag{F.3}$$

Z. Wei et al., *Energy Efficient Full Duplex Wireless Communication Systems*,
Wireless Networks, https://doi.org/10.1007/978-981-97-8915-3

Then the objective function can be transformed into

$$\underset{T_1,T_2,\boldsymbol{p_s},\boldsymbol{p_r},\boldsymbol{\beta}}{\operatorname{argmin}} \frac{1}{2}\Bigg(\frac{1}{\sum_{k=1}^{K}\sum_{n=1}^{N}\beta_{k,n}log_2(1+\frac{p_{s,k,n}\gamma_{sr,k,n}}{\beta_{k,n}})}+ \frac{1}{\sum_{k=1}^{K}\sum_{n=1}^{N}\beta_{k,n}log_2(1+\frac{p_{r,k,n}\gamma_{rd,k,n}}{\beta_{k,n}})}\Bigg) \tag{F.4}$$

For ease of notation, we write the two inter functions as

$$u=\sum_{k=1}^{K}\sum_{n=1}^{N}\beta_{k,n}log_2(1+\frac{p_{s,k,n}\gamma_{sr,k,n}}{\beta_{k,n}}), \tag{F.5}$$

and

$$v=\sum_{k=1}^{K}\sum_{n=1}^{N}\beta_{k,n}log_2(1+\frac{p_{r,k,n}\gamma_{rd,k,n}}{\beta_{k,n}}). \tag{F.6}$$

The Hessian matrix of function u is given as

$$\boldsymbol{H}\big(u(T_1,T_2,\boldsymbol{p_s},\boldsymbol{p_r},\boldsymbol{\beta})\big)=\frac{p_{s,k,n}\gamma_{sr,k,n}^2}{(\beta_{k,n}+p_{s,k,n}\gamma_{sr,k,n})^2}\begin{pmatrix}-\frac{p_{s,k,n}}{\beta_{k,n}} & 1\\ 1 & -\frac{\beta_{k,n}}{p_{s,k,n}}\end{pmatrix}. \tag{F.7}$$

It is obvious that (F.7) is a negative semi-definite matrix, meaning that u is a concave function. Also, the Hessian matrix of function v is given as

$$\boldsymbol{H}\big(v(T_1,T_2,\boldsymbol{p_s},\boldsymbol{p_r},\boldsymbol{\beta})\big)=\frac{p_{r,k,n}\gamma_{rd,k,n}^2}{(\beta_{k,n}+p_{r,k,n}\gamma_{rd,k,n})^2}\begin{pmatrix}-\frac{p_{r,k,n}}{\beta_{k,n}} & 1\\ 1 & -\frac{\beta_{k,n}}{p_{r,k,n}}\end{pmatrix}, \tag{F.8}$$

which is also a negative semi-definite matrix, indicating that v is a concave function. Since functions u and v are both concave and positive, it is concluded that the equivalent function $\frac{1}{2}(\frac{1}{u}+\frac{1}{v})$ is a convex function with one and only one optimal solution in the considered domain.

References

1. D. Sabella, A. de Domenico, E. Katranaras, M. A. Imran, M. di Girolamo, U. Salim, M. Lalam, K. Samdanis, A. Maeder, Energy efficiency benefits of ran-as-a-service concept for a cloud-based 5g mobile network infrastructure. IEEE Access **2**, 1586–1597 (2014)
2. D.W.K. Ng, E.S. Lo, R. Schober, Dynamic resource allocation in MIMO-OFDMA systems with full-duplex and hybrid relaying. IEEE Trans. Commun. **60**(5), 1291–1304 (2012)
3. M. Duarte, A. Sabharwal, V. Aggarwal, R. Jana, K.K. Ramakrishnan, C.W. Rice, N.K. Shankaranarayanan, Design and characterization of a full-duplex multiantenna system for wifi networks. IEEE Trans. Veh. Technol. **63**(3), 1160–1177 (2014)
4. Z. Wei, X. Zhu, S. Sun, Y. Huang, L. Dong, Y. Jiang, Full-duplex versus half-duplex amplify-and-forward relaying: which is more energy efficient in 60-GHZ dual-hop indoor wireless systems? IEEE J. Sel. Areas Commun. **33**(12), 2936–2947 (2015)
5. M. Jain, Practical, real-time, full duplex, in *Proceeding of 2011 ACM International Conference on Mobile Computing and Networking, HK, China* (2011), pp. 301–312
6. H. Ju, E. Oh, D. Hong, Improving efficiency of resource usage in two-hop full duplex relay systems based on resource sharing and interference cancellation. IEEE Trans. Wireless Commun. **8**(8), 3933–3938 (2009)
7. I. Krikidis, H.A. Suraweera, P.J. Smith, C. Yuen, Full-duplex relay selection for amplify-and-forward cooperative networks. IEEE Trans. Wireless Commun. **11**(12), 4381–4393 (2012)
8. H. Cui, M. Ma, L. Song, B. Jiao, Relay selection for two-way full duplex relay networks with amplify-and-forward protocol. IEEE Trans. Wireless Commun. **13**(7), 3768–3777 (2014)
9. A. Ikhlef, J. Kim, R. Schober, Mimicking full-duplex relaying using half-duplex relays with buffers. IEEE Trans. Veh. Technol. **61**(7), 3025–3037 (2012)
10. S. Huberman, T. Le-Ngoc, Mimo full-duplex precoding: a joint beamforming and self-interference cancellation structure. IEEE Trans. Wireless Commun. **14**(4), 2205–2217 (2015)
11. E. Everett, A. Sahai, A. Sabharwal, Passive self-interference suppression for full-duplex infrastructure nodes. IEEE Trans. Wireless Commun. **12**(2), 680–694 (2014)
12. V. Syrjala, M. Valkama, L. Anttila, T. Riihonen, D. Korpi, Analysis of oscillator phase-noise effects on self-interference cancellation in full-duplex ofdm radio transceivers. IEEE Trans. Wireless Commun. **13**(6), 2977–2990 (2014)
13. Z. He, S. Shao, Y. Shen, C. Qing, Y. Tang, Performance analysis of RF self-interference cancellation in full-duplex wireless communications. IEEE Wireless Commun. Lett. **3**(4), 405–408 (2014)
14. B. François, P. Reynaert, A fully integrated transformer-coupled power detector with 5 GHz RF PA for WLAN 802.11ac in 40 nm CMOS. IEEE J. Solid-State Circuits **50**(5), 1237–1250 (2015)

Z. Wei et al., *Energy Efficient Full Duplex Wireless Communication Systems*, Wireless Networks, https://doi.org/10.1007/978-981-97-8915-3

15. K. Yamamoto, T. Moriwaki, T. Fujii, J. Otsuji, M. Miyashita, Y. Miyazaki, K. Nishitani, A 2.2-v operation, 2.4-GHz single-chip GaAs MMIC transceiver for wireless applications. IEEE J. Solid-State Circuits **34**(4), 502–512 (1999)
16. A. Siligaris, Y. Hamada, C. Mounet, C. Raynaud, B. Martineau, N. Deparis, N. Rolland, M. Fukaishi, P. Vincent, A 60 GHz power amplifier with 14.5 DBM saturation power and 25% peak PAE in CMOS 65 nm SOI. IEEE J. Solid-State Circuits **45**(7), 1286–1294 (2010)
17. C. Marcu, D. Chowdhury, C. Thakkar, J.D. Park, L.K. Kong, M. Tabesh, Y. Wang, B. Afshar, A. Gupta, A. Arbabian, S. Gambini, R. Zamani, E. Alon, A.M. Niknejad, A 90 nm CMOS low-power 60 GHz transceiver with integrated baseband circuitry. IEEE J. Solid-State Circuits **44**(12), 3434–3447 (2009)
18. X. Zhu, R.D. Murch, Layered space-frequency equalization in a single-carrier mimo system for frequency-selective channels. IEEE Trans. Wireless Commun. **3**(3), 701–708 (2004)
19. H. Xie, F. Gao, S. Zhang, S. Jin, A unified transmission strategy for TDD/FDD massive MIMO systems with spatial basis expansion model. IEEE Trans. Veh. Technol. **66**(4), 3170–3184 (2017)
20. Z. Wei, X. Zhu, S. Sun, Y. Huang, Energy-efficiency-oriented cross-layer resource allocation for multiuser full-duplex decode-and-forward indoor relay systems at 60 GHz. IEEE J. Sel. Areas Commun. **34**(12), 3366–3379 (2016)
21. B. Radunovic, D. Gunawardena, P. Key, A. Proutiere, N. Singh, V. Balan, G. Dejean, Rethinking indoor wireless mesh design: low power, low frequency, full-duplex, in *Proceeding of 2010 IEEE International Workshop on Wireless Mesh Networks, Boston, USA* (2010), pp. 1–6
22. A. Almradi, K.A. Hamdi, On the outage probability of MIMO full-duplex relaying: impact of antenna correlation and imperfect CSI. IEEE Trans. Veh. Technol. **66**(5), 3957–3965 (2017)
23. A.C. Cirik, Y. Rong, Y. Hua, Achievable rates of full-duplex MIMO radios in fast fading channels with imperfect channel estimation. IEEE Trans. Signal Process. **62**(15), 3874–3886 (2014)
24. D. Nguyen, L.N. Tran, P. Pirinen, M. Latva-aho, On the spectral efficiency of full-duplex small cell wireless systems. IEEE Trans. Wireless Commun. **12**(9), 4896–4910 (2014)
25. B.P. Day, A.R. Margetts, D.W. Bliss, P. Schniter, Full-duplex bidirectional MIMO: achievable rates under limited dynamic range. IEEE Trans. Signal Process. **60**(7), 3702–3713 (2012)
26. C. Xiong, G.Y. Li, S. Zhang, Y. Chen, S. Xu, Energy- and spectral-efficiency tradeoff in downlink OFDMA networks. IEEE Trans. Wireless Commun. **10**(11), 3874–3886 (2011)
27. C. Xiong, G.Y. Li, Y. Liu, Y. Chen, S. Xu, Energy-efficient design for downlink OFDMA with delay-sensitive traffic. IEEE Trans. Wireless Commun. **12**(6), 3085–3095 (2013)
28. D.W.K. Ng, E.S. Lo, R. Schober, Energy-efficient resource allocation in OFDMA systems with large numbers of base station antennas. IEEE Trans. Wireless Commun. **11**(9), 3292–3304 (2012)
29. H.A. Suraweera, I. Krikidis, G. Zheng, C. Yuen, P.J. Smith, Low-complexity end-to-end performance optimization in MIMO full-duplex relay systems. IEEE Trans. Wireless Commun. **13**(2), 913–927 (2014)
30. T. Riihonen, S. Werner, R. Wichman, Mitigation of loopback self-interference in full-duplex MIMO relays. IEEE Trans. Signal Process. **59**(12), 5983–5993 (2011)
31. G. Amarasuriya, C. Tellambura, M. Ardakani, Two-way amplify-and-forward multiple-input multiple-output relay networks with antenna selection. IEEE J. Sel. Areas Commun. **30**(8), 1513–1529 (2012)
32. J.H. Lee, O.S. Shin, Full-duplex relay based on distributed beamforming in multiuser MIMO systems. IEEE Trans. Veh. Technol. **62**(4), 1855–1860 (2013)
33. G. Amarasuriya, C. Tellambura, M. Ardakani, Multi-way MIMO amplify-and-forward relay networks with zero-forcing transmission. IEEE Trans. Commun. **61**(12), 4847–4863 (2013)
34. H.Q. Ngo, H.A. Suraweera, M. Matthaiou, E.G. Larsson, Multipair full-duplex relaying with massive arrays and linear processing. IEEE J. Sel. Areas Commun. **32**(9), 1721–1737 (2014)
35. B. Yin, M. Wu, C. Studer, J.R. Cavallaro, J. Lilleberg, Full-duplex in large-scale wireless systems, in *Proceeding of 2013 Asilomar International Conference on Signals, Systems and Computers, CA, USA* (2013), pp. 1623–1627

36. X. Gao, O. Edfors, F. Tufvesson, E.G. Larsson, Massive MIMO in real propagation environments: do all antennas contribute equally? IEEE Trans. Commun. **63**(11), 3917–3928 (2015)
37. J. Joung, Y.K. Chia, S. Sun, Energy-efficient, large-scale distributed-antenna system (L-DAS) for multiple users. IEEE J. Sel. Top. Signal Process. **8**(5), 954–965 (2014)
38. D.W.K. Ng, Y. Wu, R. Schober, Power efficient resource allocation for full-duplex radio distributed antenna networks. IEEE Trans. Wireless Commun. **15**(4), 2896–2911 (2016)
39. R. Bolla, R. Bruschi, F. Davoli, F. Cucchietti, Energy efficiency in the future internet: a survey of existing approaches and trends in energy-aware fixed network infrastructures. IEEE Commun. Surv. Tuts. **13**(2), 223–244 (2011)
40. G. Liu, F.R. Yu, H. Ji, V.C.M. Leung, X. Li, In-band full-duplex relaying: a survey, research issues and challenges. IEEE Commun. Surv. Tuts. **17**(2), 500–524 (2015)
41. L. Lu, G.Y. Li, A.L. Swindlehurst, A. Ashikhmin, R. Zhang, An overview of massive MIMO: benefits and challenges. IEEE J. Sel. Top. Signal Process. **8**(5), 742–758 (2014)
42. S. Cui, A.J. Goldsmith, A. Bahai, Energy-constrained modulation optimization. IEEE Trans. Wireless Commun. **4**(5), 2349–2360 (2005)
43. D. Nguyen, L.N. Tran, P. Pirinen, M. Latva-aho, Precoding for full duplex multiuser mimo systems: spectral and energy efficiency maximization. IEEE Trans. Signal Process. **61**(16), 4038–4050 (2013)
44. X. Xiao, X. Tao, J. Lu, Energy-efficient resource allocation in LTE-based Mimo-OFDMA systems with user rate constraints. IEEE Trans. Veh. Technol. **64**(1), 185–197 (2015)
45. Y. Sun, D.W.K. Ng, J. Zhu, R. Schober, Multi-objective optimization for robust power efficient and secure full-duplex wireless communication systems. IEEE Trans. Wireless Commun. **15**(8), 5511–5526 (2016)
46. M. Jaggi, Revisiting frank-wolfe projection-free sparse convex optimization, in *Proceeding of 2013 IEEE International Conference on Machine Learning, Atlanta, USA* (2013), pp. 427–435
47. Y. Cheng, M. Pesavento, A. Philipp, Joint network optimization and downlink beamforming for comp transmissions using mixed integer conic programming. IEEE Trans. Signal Process. **61**(16), pp. 3972–3987 (2013)
48. I. Krikidis, S. Timotheou, S. Nikolaou, G. Zheng, D.W.K. Ng, R. Schober, Simultaneous wireless information and power transfer in modern communication systems. IEEE Commun. Mag. **52**(11), 104–110 (2014)
49. J. Wang, C. Jiang, Z. Han, Y. Ren, L. Hanzo, Network association strategies for an energy harvesting aided super-wifi network relying on measured solar activity. IEEE J. Sel. Areas Commun. **34**(12), 3785–3797 (2016)
50. Q. Li, Q. Zhang, J. Qin, Secure relay beamforming for simultaneous wireless information and power transfer in nonregenerative relay networks. IEEE Trans. Veh. Technol. **63**(5), 2462–2467 (2014)
51. A.A. Nasir, X. Zhou, S. Durrani, R.A. Kennedy, Relaying protocols for wireless energy harvesting and information processing. IEEE Trans. Wireless Commun. **12**(7), 3622–3636 (2013)
52. Z. Ding, I. Krikidis, B. Sharif, H.V. Poor, Wireless information and power transfer in cooperative networks with spatially random relays. IEEE Trans. Wireless Commun. **13**(8), 4440–4453 (2014)
53. G. Amarasuriya, E.G. Larsson, H.V. Poor, Wireless information and power transfer in multiway massive MIMO relay networks. IEEE Trans. Wireless Commun. **15**(6), 3837–3855 (2016)
54. K.H. Liu, Performance analysis of relay selection for cooperative relays based on wireless power transfer with finite energy storage. IEEE Trans. Veh. Technol. **65**(7), 5110–5121 (2015)
55. Z. Yang, Z. Ding, P. Fan, G.K. Karagiannidis, Outage performance of cognitive relay networks with wireless information and power transfer. IEEE Trans. Veh. Technol. **65**(5), 3828–3833 (2015)
56. H. Lee, C. Song, S.H. Choi, I. Lee, Outage probability analysis and power splitter designs for swipt relaying systems with direct link. IEEE Commun. Lett. **21**(3), 648–651 (2016)

57. D.W.K. Ng, E.S. Lo, R. Schober, Wireless information and power transfer: energy efficiency optimization in OFDMA systems. IEEE Trans. Wireless Commun. **12**(12), 6352–6370 (2013)
58. C. Zhong, H.A. Suraweera, G. Zheng, I. Krikidis, Z. Zhang, Wireless information and power transfer with full duplex relaying. IEEE Trans. Commun. **62**(10), 3447–3461 (2014)
59. Z. Wen, X. Liu, N.C. Beaulieu, R. Wang, S. Wang, Joint source and relay beamforming design for full-duplex MIMO AF relay swipt systems. IEEE Commun. Lett. **20**(2), 320–323 (2016)
60. M. Mohammadi, B.K. Chalise, H.A. Suraweera, C. Zhong, G. Zheng, I. Krikidis, Throughput analysis and optimization of wireless-powered multiple antenna full-duplex relay systems. IEEE Trans. Commun. **64**(4), 1769–1785 (2016)
61. H. Ju, R. Zhang, Optimal resource allocation in full-duplex wireless-powered communication network. IEEE Trans. Commun. **62**(10), 3528–3540 (2014)
62. X. Kang, C.K. Ho, S. Sun, Full-duplex wireless-powered communication network with energy causality. IEEE Trans. Wireless Commun. **14**(10), 5539–5551 (2015)
63. V.D. Nguyen, T.Q. Duong, H.D. Tuan, O.S. Shin, H.V. Poor, Spectral and energy efficiencies in full-duplex wireless information and power transfer. IEEE Trans. Commun. **65**(5), 2220–2233 (2017)
64. Y. Zeng, R. Zhang, Optimized training design for wireless energy transfer. IEEE Trans. Commun. **63**(2), 536–550 (2015)
65. R. Tao, A. Salem, K.A. Hamdi, Adaptive relaying protocol for wireless power transfer and information processing. IEEE Commun. Lett. **20**(10), 2027–2030 (2016)
66. Y. Zeng, H. Chen, R. Zhang, Bidirectional wireless information and power transfer with a helping relay. IEEE Commun. Lett. **20**(5), 862–865 (2016)
67. Z. Zhou, M. Peng, Z. Zhao, Y. Li, Joint power splitting and antenna selection in energy harvesting relay channels. IEEE Signal Process. Lett. **22**(7), 823–827 (2014)
68. S. Lohani, R.A. Loodaricheh, E. Hossain, V.K. Bhargava, On multiuser resource allocation in relay-based wireless-powered uplink cellular networks. IEEE Trans. Wireless Commun. **15**(3), 1851–1865 (2015)
69. D.W.K. Ng, E.S. Lo, R. Schober, Robust beamforming for secure communication in systems with wireless information and power transfer. IEEE Trans. Wireless Commun. **13**(8), 4599–4615 (2014)
70. R. Feng, Q. Li, Q. Zhang, J. Qin, Robust secure transmission in miso simultaneous wireless information and power transfer system. IEEE Trans. Veh. Technol. **64**(1), 400–405 (2014)
71. J. Xu, L. Liu, R. Zhang, Multiuser miso beamforming for simultaneous wireless information and power transfer. IEEE Trans. Signal Process. **62**(18), 4798–4810 (2014)
72. F. Shan, J. Luo, W. Wu, X. Shen, Optimal wireless power transfer scheduling for delay minimization, in *IEEE INFOCOM 2016-The 35th Annual IEEE International Conference on Computer Communications* (IEEE, Piscataway, 2016), pp. 1–9
73. D. Hwang, K.C. Hwang, D.I. Kim, T.J. Lee, Self-energy recycling for RF powered multi-antenna relay channels. IEEE Trans. Wireless Commun. **16**(2), 812–824 (2017)
74. T. Riihonen, L. Zhao, M. Vehkaperä, X. Wang, On the feasibility of full-duplex relaying powered by wireless energy transfer, in *2016 IEEE 17th International Workshop on Signal Processing Advances in Wireless Communications (SPAWC)* (IEEE, Piscataway, 2016), pp. 1–5
75. Z. Hu, C. Yuan, F. Zhu, F. Gao, Weighted sum transmit power minimization for full-duplex system with swipt and self-energy recycling. IEEE Access **4**, 4874–4881 (2016)
76. Y. Zeng, R. Zhang, Full-duplex wireless-powered relay with self-energy recycling. IEEE Wireless Commun. Lett. **4**(2), 201–204 (2015)
77. Y. Wang, R. Sun, X. Wang, Transceiver design to maximize the weighted sum secrecy rate in full-duplex swipt systems. IEEE Signal Process. Lett. **23**(6), 883–887 (2016)
78. N. Zhou, Dynamic resource management for next generation broadband wireless communication systems. Ph.D. Thesis, The University of Liverpool, Liverpool (2010)
79. M. Maso, C.F. Liu, C.H. Lee, T.Q.S. Quek, L.S. Cardoso, Energy-recycling full-duplex radios for next-generation networks. IEEE J. Sel. Areas Commun. **33**(12), 2948–2962 (2015)

80. Z. Wei, X. Zhu, S. Sun, Y. Huang, A. Al-Tahmeesschi, Y. Jiang, Energy-efficiency of millimeter-wave full-duplex relaying systems: challenges and solutions. IEEE Access **4**, 4848–4860 (2016)
81. Y. Liu, Z. Qin, M. Elkashlan, Z. Ding, A. Nallanathan, L. Hanzo, Nonorthogonal multiple access for 5G and beyond. Proc. IEEE **105**(12), 2347–2381 (2017)
82. C.L. Wang, J.Y. Chen, Y.J. Chen, Power allocation for a downlink non-orthogonal multiple access system. IEEE Wireless Commun. Lett. **5**(5), 532–535 (2016)
83. Z. Yang, Z. Ding, P. Fan, Z. Ma, Outage performance for dynamic power allocation in hybrid non-orthogonal multiple access systems. IEEE Commun. Lett. **20**(8), 1695–1698 (2016)
84. Z. Yang, Z. Ding, P. Fan, G.K. Karagiannidis, On the performance of non-orthogonal multiple access systems with partial channel information. IEEE Trans. Commun. **64**(2), 654–667 (2015)
85. L. Lei, D. Yuan, C.K. Ho, S. Sun, Power and channel allocation for non-orthogonal multiple access in 5G systems: tractability and computation. IEEE Trans. Wireless Commun. **15**(12), 8580–8594 (2016)
86. S. Timotheou, I. Krikidis, Fairness for non-orthogonal multiple access in 5G systems. IEEE Signal Process. Lett. **22**(10), 1647–1651 (2015)
87. Y. Liu, M. Elkashlan, Z. Ding, G.K. Karagiannidis, Fairness of user clustering in MIMO non-orthogonal multiple access systems. IEEE Commun. Lett. **20**(7), 1465–1468 (2016)
88. Z. Ding, F. Adachi, H.V. Poor, The application of MIMO to non-orthogonal multiple access. IEEE Trans. Wireless Commun. **15**(1), 537–552 (2015)
89. Z. Ding, R. Schober, H.V. Poor, A general MIMO framework for NOMA downlink and uplink transmission based on signal alignment. IEEE Trans. Wireless Commun. **15**(6), 4438–4454 (2016)
90. Z. Ding, P. Fan, H. V. Poor, Impact of user pairing on 5G nonorthogonal multiple-access downlink transmissions. IEEE Trans. Veh. Technol. **65**(8), 6010–6023 (2015)
91. Y. Liu, Z. Ding, M. Elkashlan, J. Yuan, Nonorthogonal multiple access in large-scale underlay cognitive radio networks. IEEE Trans. Veh. Technol. **65**(12), 10152–10157 (2016)
92. J.B. Kim, I.H. Lee, Capacity analysis of cooperative relaying systems using non-orthogonal multiple access. IEEE Commun. Lett. **19**(11), 1949–1952 (2015)
93. Z. Ding, M. Peng, H.V. Poor, Cooperative non-orthogonal multiple access in 5G systems. IEEE Commun. Lett. **19**(8), 1462–1465 (2015)
94. C. Zhong, Z. Zhang, Non-orthogonal multiple access with cooperative full-duplex relaying. IEEE Commun. Lett. **20**(12), 2478–2481 (2016)
95. Z. Zhang, Z. Ma, M. Xiao, Z. Ding, P. Fan, Full-duplex device-to-device-aided cooperative nonorthogonal multiple access. IEEE Trans. Veh. Technol. **66**(5), 4467–4471 (2016)
96. L. Lv, J. Chen, Q. Ni, Cooperative non-orthogonal multiple access in cognitive radio. IEEE Commun. Lett. **20**(10), 2059–2062 (2016)
97. G. Liu, X. Chen, Z. Ding, Z. Ma, F.R. Yu, Hybrid half-duplex/full-duplex cooperative non-orthogonal multiple access with transmit power adaptation. IEEE Trans. Wireless Commun. **17**(1), 506–519 (2017)
98. L. Zhang, J. Liu, M. Xiao, G. Wu, Y.C. Liang, S. Li, Performance analysis and optimization in downlink NOMA systems with cooperative full-duplex relaying. IEEE J. Sel. Areas Commun. **35**(10), 2398–2412 (2017)
99. F. Fang, H. Zhang, J. Cheng, V.C. Leung, Energy-efficient resource allocation for downlink non-orthogonal multiple access network. IEEE Trans. Commun. **64**(9), 3722–3732 (2016)
100. J. Choi, Minimum power multicast beamforming with superposition coding for multiresolution broadcast and application to Noma systems. IEEE Trans. Commun. **63**(3), 791–800 (2015)
101. B. Klaiqi, X. Chu, J. Zhang, Energy-efficient and low signaling overhead cooperative relaying with proactive relay subset selection. IEEE Trans. Commun. **64**(3), 1001–1015 (2016)
102. Y. Xiao, L.J. Cimini, Impact of overhead on spectral efficiency of cooperative relaying. IEEE Trans. Wireless Commun. **12**(5), 2228–2239 (2013)

The manufacturer's authorised representative in the EU is Springer Nature Customer Service Centre GmbH, Europaplatz 3, 69115 Heidelberg, Germany. If you have any concerns regarding our products, please contact ProductSafety@springernature.com

Printed and bound by CPI Group (UK) Ltd, Croydon, CR0 4YY
01/10/2026
02225916-0001